Fernando Pacheco Torgal
Said Jalali

Eco-efficient Construction and Building Materials

Fernando Pacheco Torgal
C-TAC Research Unit
University of Minho
Guimarães
Portugal
e-mail: torgal@civil.uminho.pt

Said Jalali
Department of Civil Engineering
University of Minho
Guimarães
Portugal
e-mail: said@civil.uminho.pt

ISBN 978-0-85729-891-1
e-ISBN 978-0-85729-892-8
DOI 10.1007/978-0-85729-892-8
Springer London Dordrecht Heidelberg New York

British Library Cataloguing in Publication Data
A catalogue record for this book is available from the British Library

Cover design: eStudio Calamar S.L.

Printed on acid-free paper

Springer is part of Springer Science+Business Media (www.springer.com)

Eco-efficient Construction and Building Materials

Contents

Chapter 1
Introduction

1.1 General

The most important environmental problem faced by Planet Earth is related to the increase of the mean air temperature (IPCC 2007; Schellnhuber 2008), which is due to the increase of carbon dioxide (CO_2) in the atmosphere. In the early eighteenth century, the concentration level of atmospheric CO_2 was 280 parts per million (ppm), at present it is already 430 ppm and growing at a pace above 2 ppm/yr. Keeping the current level of emissions (which is unlikely given the high economic growth of less developed countries with consequent increases in emission rates) will imply a CO_2 concentration of 550 ppm in the year 2050 (Stern 2006). The rise in the mean air temperature will lead to a rise in the sea level caused by thermal expansion of the water. Until 2100 it is expected that the sea level will rise, between 0.18 and 0.59 m (Meehl et al. 2007). When the sea level rises above 0.40 m it will submerge 11% of the area of Bangladesh and as a result of this fact will lead to almost 10 million homeless (IPCC 2007). This rise does not include the melting of the ice caps, whose impacts are not quantified accurately and can be very substantial, meaning a rise in sea level of almost seven meters (Broecker and Kunzig 2008). Another consequence of the increase in the mean air temperature is the occurrence of increasingly extreme atmospheric events. Not only long-term dry periods that could enhance the action of fires, but also heavy rains and even hurricanes (Allan and Soden 2008; Liu et al. 2009; Zolina et al. 2010). Saunders and Lea (2008) mentioned that between 1996 and 2005 the occurrence of hurricanes, increased 40% due to an increase of only 0.5°C in the temperature of seawater. The rise in the seawater temperature will eventually stop the thermohaline circulation (Fig. 1.1), which is related to the movement of ocean mass due to its salinity and temperature (together with the action of the wind), being responsible for carrying heat from the tropics to areas of higher latitudes.

In the Polar regions the water becomes denser and sinks moving back to the South. The thermohaline circulation in the North Atlantic is responsible for the fact that the climate on the West coast of Europe is more moderate than in other

F. P. Torgal and S. Jalali, *Eco-efficient Construction and Building Materials*,
DOI: 10.1007/978-0-85729-892-8_1,

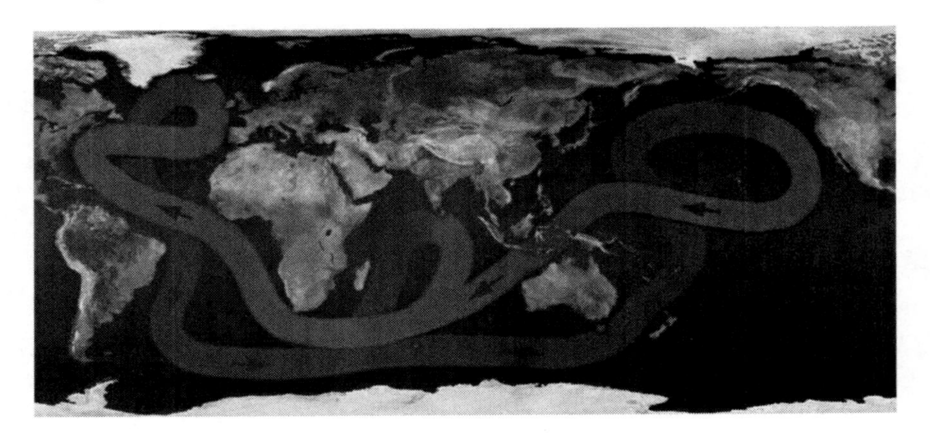

Fig. 1.1 Thermohaline circulation

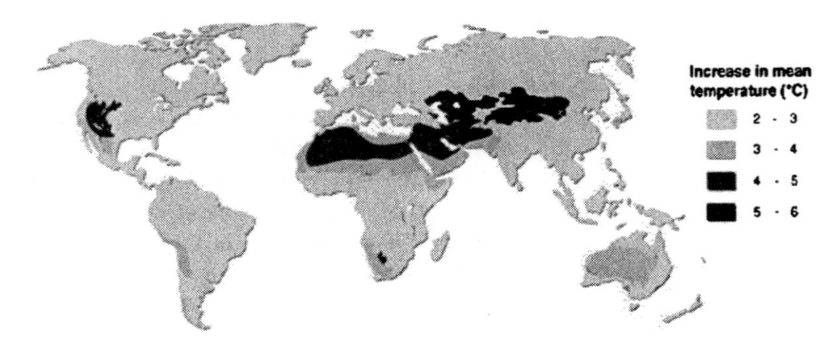

Fig. 1.2 Increase in desert areas according to the increase in the mean air temperature (UNEP 2006)

areas located on the same latitude (Vellinga and Wood 2002), it is also responsible for the increase in the level of oxygen at the bottom of the ocean. If the thermohaline circulation stops it will not only lead to extreme weather events, but will also lead to serious changes in the oceanic biodiversity (Meehl et al. 2007; Vellinga and Wood 2008). The rise in the mean air temperature may also lead to the thawing of the permafrost (permanently frozen ground), where approx. 1×10^6 million tons (1,000 giga-tones) of CO_2 are still retained. In order to comprehend this number we should remember that the atmosphere already contains 0.7 million tonnes $\times$ 10^6 (750 giga-tones) of CO_2 and that the human activity produces annually 6.5 gigatons of CO_2 (Bourne 2008). Some projections show that the rise in the mean air temperature will also lead to an increase in desert-like areas (Fig. 1.2).

The majority of CO_2 emissions come from burning fossil fuels for energy production. Coal plants are responsible for 20% of CO_2 emissions worldwide and it is known that China, as the world's largest producer of coal, puts into operation a new coal plant every two weeks. Chinese coal plants are responsible for 80% of electricity generation (Shealy and Dorian 2009). One of the most obvious and immediate consequences of burning coal, is the fact that 16 of the 20 most polluted cities of the Earth are located in China, in addition, the World Bank estimates that every year 400,000 people die in China, because of poor air quality (Wang and Watson 2010). China recently surpassed the United States in terms of global CO_2 emissions; however, this scenario can get worse because China still has a very low level of CO_2 emissions per capita (5 ton.), when compared to the emissions of United States (20 ton.). Since the increase in the world population that is expected to rise from the current 5,600 to 7,900 million inhabitants in 2050 (UN 2008) will occur in countries with low CO_2 emissions per capita, this does not constitute an optimistic scenario in terms of agreements to reduce overall CO_2 emissions. It is then understandable why the countries present in the Copenhagen Climate Summit (2009) failed to reach a binding and ambitious agreement towards the reduction of CO_2 emissions (Goldenberg and Prado 2010). Despite this lack of agreement it is inevitable that in a short period all countries must lower their CO_2 emissions and the most industrialized countries should do it more significantly.

Another serious environmental problem relates to the loss of biodiversity caused by human activity. The Convention on Biological Diversity, an organization founded in 1993, defines biodiversity as "the diversity among living organisms, whatever their origin, including terrestrial, marine and other aquatic ecosystems and the ecological complexes of which they are part, which includes diversity within species, between species and of ecosystems". No one knows for sure how many species exist on Planet Earth. According to the report "Global Biodiversity Outlook", of the Convention on Biological Diversity, 1.75 million species have already been identified. However, it is estimated that their number is much higher, approaching 15 million. The current rate of species extinction varies between 1,000 and 10,000 times higher than the average extinction paleontology rate. In Europe alone, 42% of mammals, 15% of birds and 45% of butterflies and reptiles, are at risk of extinction. The climate change, the high rates of urbanization, the excessive exploitation of resources and the consequent production of waste, are high-risk factors for the preservation of biodiversity:

- Mankind already uses almost 50% of freshwater reserves and the scenarios for the increase in the world population, create a serious problem.
- Almost 50% of world grain is being fed to livestock rather than being consumed directly by humans.
- Agriculture, cattle and other livestock consume 70% of the freshwater reserves.
- Approximately 24% of the land is already cultivated with some plant species (Millenium 2005).
- The use of water for cotton irrigation purposes led to a decrease of 74% of the Aral Sea area, which once would have been the fourth largest lake in the world (Micklin 2007).

Fig. 1.3 Albatross corpse with a stomach full of plastic waste

- Between 1960 and 1990 the use of fertilizers increased by 300% (Millenium 2005). The major part of fertilizers used in agricultural processes are dragged into lakes, rivers and into the sea, contributing to eutrophication (Spiertz 2010).
- Currently only 12% of the soils and 0.5% of the seas are subject to conservation measures.
- Over the past 300 years there has been a 40% reduction in the forest area (Ring et al. 2010).
- Each year 13.7 million hectares of forest are thinned. This results in a negative net balance of 7.3 million hectares per year (Gore 2009).
- Almost 20% of coral reefs have been destroyed and another 20% are at risk (Millenium 2005).
- Transportation accounts for 26% of global carbon dioxide emissions, and existing projects indicate an increase in emissions in this sector (Chapman 2007).
- Between 1960 and 2000 the production of plastic resins increased 25-fold, while reused plastic only grew by 5% (Derraik 2002).
- In the U.S. alone 50,000 million plastic bottles are landfilled (Gore 2009).
- Between 500,000 million to one billion plastic bags are used each year, some of which end up as waste in the oceans.

Moore et al. (2001) mention a deposit of plastic waste floating in the Pacific Ocean with a diameter of about 1,000 km and almost 3 million tons known as the "The Great Pacific garbage patch". According to Moore (2008) most of these plastics, among others, are eaten by turtles, fish and seabirds (Fig. 1.3).

Beyond the ethical arguments related to the importance of the intrinsic value of all species, biodiversity is a guarantee of water and air purification, production of food resources and other products such as vaccines, antibiotics etc. For instance, approximately 80% of the major crops benefit from the action of natural pollinators like bees. Unfortunately the deaths of large amounts of swarms have become frequent (Genersch 2010). Part of the explanation for this phenomenon is related to the high amount of pesticides used in agricultural production (Brittain et al. 2010). The increase in CO_2 emissions will also lead to the acidification of sea water, with negative consequences in coral reefs putting habitats of high economic value at risk (Anthony et al. 2008). Coral reef habitats represent fish resources that feed more than 1,000 million people and have an economic value estimated at 20,000 million € (Bourne 2008). According to Costanza et al. (1998), the services provided free of charge by Nature reach almost 33 billion (10^{12}) dollars/year. As a comparison the global GDP in the same period was 18 billion (10^{12}) dollars per year, roughly half the value of the services and products provided by Nature. Balmford et al. (2002) studied the economic benefits of the protection of areas subject to conservation measures, pointing to a gain of 1 to 100. A study begun in 2007 under the G8 and five emerging economies (Brazil, China, India, Mexico and South Africa), shows that the investment in the protection of ecosystems, may have a return between 25 and 100 times the amount invested (Ring et al. 2010). For instance the expansion of marine protected areas from the current 0.5% to 30%, would generate a benefit between 40 and 50 billion dollars.

1.2 Sustainable Development

The report "Our common future", also known as the Bruntland (1987) report, is where for the first time appears the expression sustainable development as one that "*meets the needs of the present without compromising the ability of future generations to meet theirs*". It is worth remembering that there are many definitions about the concept of sustainability, which are not as simplistic as the one present in the Brundtland report (Robinson 2004; Destatte 2010). Pearce and Walrath (2008) presented a list of over 160 definitions of sustainability. Other authors (Dietz et al. 2009) suggest that the concept of sustainability should focus on balancing the welfare of humankind and the environmental impacts arising from it. Some authors (Clayton 2001; Choi and Pattent 2001) argue that "sustainable development" is an oxymoron, because we cannot have development/growth for the entire world population, and at the same time expect that this development may be compatible with environmental sustainability. This view is not without merit because when we look to the ecologic footprint, the concept developed by Rees and Wackernagel (1966) to measure the world biocapacity, we realize that we are already living beyond the Earth's biocapacity (Fig. 1.4).

Concerns about sustainable development were not born with the Brundtland Report. Table 1.1 presents the most relevant events in the context of sustainable development. The publication of the book "*Silent Spring*" which warned about the harmful effects of pesticides can be considered an important landmark in this context and similarly with the fourth IPCC report which showed that climate change is a fact (not just a theory) related to the emission of greenhouse gases (GHGs). As for the Kyoto Protocol (UNFCCC 1997), it represents the most known instrument for the reduction of green house gases (GHGs). The signatory countries

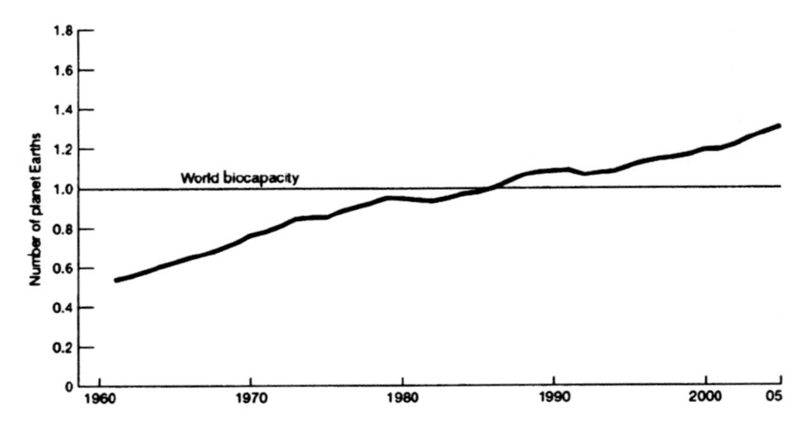

Fig. 1.4 Global ecological footprint, 1961–2005 (WWF 2008; Wiedmann and Barrett 2010)

Table 1.1 Landmark events related to sustainable development (adapted from Mateus 2009)

Year	Fact
1962	Publication of the book "*Silent Spring*" by the biologist Rachel Carson
1972	Presentation of the Report "*The Limits of Grow*" by the Club of Rome
	UN Conference in Stockholm about the Human Environment which led to the appearance of the UN Environment Program—UNEP
1979	Bern Convention on Habitats
	Geneva Convention on indoor air quality
1980	The IUCN, UNEP, WWF and the UNESCO present a strategic document about the conservation of Nature
	Presentation of the Report Global 2000
1983	UN protocol about air pollution (Helsinki)
	UN comission on environment and development
1987	UN protocol about substances that damage the ozone layer (Montreal)
	Bruntland Report
1990	EU Report on urban environment
1992	Earth Summit in Rio
1997	Kyoto Summit about the global warming
2007	4th IPCC Report
	Al Gore features the movie "*An inconvenient truth*"
	The IPCC and Al Gore received the Nobel Peace prize
2009	Copenhagen Summit about climate change

(almost 200 so far) committed themselves to reduce their emission of GHGs to 5.2% by 2012 below the level of the GHG emissions in the reference year of 1990.

As for the Copenhagen Climate Summit it failed to generate a comprehensive agreement that could have a significant impact on reducing carbon dioxide emissions. Therefore different countries decide to pursue different reduction targets. The European Union agreed to reduce its overall emissions by 20% in the year 2020 with reference to the year 1990. The United States agreed to reduce its overall emissions by 17% in 2010, with reference to the year 2005. China and India did not accept a reduction in their total emissions, but rather a reduction in their carbon dioxide intensity (carbon dioxide/unit of wealth) until 2020, between 40 and 45% for China and between 20 and 25% for India. Goldenberg and Prado (2010), reviewed the above cited goals and report that they simply follow the standard "*business as usual*" for the period 1990–2007, which is clearly insufficient to achieve significant reductions by the year 2020. This is not surprising, since Stern (2006) stated the need for minimum reductions of 25% to achieve the stabilization of carbon dioxide concentration in the atmosphere.

1.3 Sustainable Construction

The construction industry is one of the largest and most active sectors throughout Europe representing 28.1% and 7.5% of employment respectively in the industry and in the European economy. With an annual turnover of 750 million euro, this sector represents 25% of all European industrial production, being the largest exporter with 52% market share. In global terms the construction industry will keep on growing at a fast pace. For instance China will need 40 billion square meters of combined residential and commercial floor space over the next 20 years—equivalent to adding one New York every two years or the area of Switzerland (Dobbs 2010). Environmentally speaking, this industry accounts for 30% of carbon dioxide emissions and the building stock consumes 42% of the energy consumed in Europe. In addition, the global construction industry consumes more raw materials (about 3000 Mt/yr, almost 50% by weight) than any other economic activity, which shows a clearly unsustainable industry. Moreover, many buildings currently suffer from problems related to excessive moisture with mold formation, or present humidity levels below 40% giving rise to respiratory diseases. Another problem affecting the quality of the indoor air has to do with the presence of construction materials with some level of toxicity, even respecting legal regulations! This aspect will be properly addressed in Chap. 2. In 1994, the International Council of Building, CIB, defined the concept of sustainable construction as the one "responsible for creating and maintaining" a healthy built environment based on the efficient use of resources and in the project based on ecological principles (Kibert 1994, 2008). Also in 1994 the CIB has defined seven principles for sustainable construction (Table 1.2).

In the previous years a number of tools have been developed aiming at the assessment of the sustainability of building stock (Table 1.3).

Some tools assess the life cycle (LCA) of buildings while others assess the sustainability of buildings or real state such as the Building Research and Consultancy's Environmental Assessment Method BREEAM-(UK, 1990) and the Leadership in Energy and Environment LEED (USA 1998), the GBTool (Canada, 1995) which later changed its name to SBTool and even the Deutsches Gütesiegel Nachhaltiges Bauen—DGNB (Germany, 2009). The first two are the most widely

Table 1.2 The principles of sustainable construction (Kibert 2008)

1	Reduce resource consumption (resources)
2	Ruse resources (reuse)
3	Use of recycle resources (recycle)
4	Protection of nature (Nature)
5	Elimination of toxics (toxics)
6	Application of life cycle costing (economics)
7	Focus on quality (quality)

Table 1.3 LCA tools for buildings (Bribian et al. 2009)

Designação	Link
ECO-QUANTUM	www.ecoquantum.nl
LEGEP	www.legep.de
EQUER	www.izuba.fr
ATHENA	www.athenasSMI.ca
OGIP	www.ogip.ch/
ECO-SOFT	www.ibo.at/de/ecosoft.htm
ENVEST 2.0	envestv2.bre.co.uk
BECOST	www.vtt.fi/rte/esitteet/ymparisto /lcahouse.html
BEES	www.bfrl.nist.gov/oae/software/bees.html
GREENCALC	www.greencalc.com
ECOEFFECT	www.ecoeffect.se
ECO-QUANTUM	www.ecoquantum.nl
LEGEP	www.legep.de
EQUER	www.izuba.fr

used and have some similarities between them (Lee and Burnett 2008; Fenner and Ryce 2008a, b). Although the tools for assessment of buildings' sustainability constitute a step forward in environmental terms, it is necessary to bear in mind that its existence entails serious disadvantages. For instance they carry the risk of portraying as sustainable options buildings that have just minor environmental improvements. Beyond the oxymoron implicit in the term "sustainable construction", as it happens for "sustainable development" (Clayton 2001; Choi and Pattent 2001) it may well be the case of sending the wrong message to the construction market and the general population, passing the idea that sustainable construction is only a matter of a slight change in building practices. Similar observations can be made about the tools aiming for the assessment of the buildings sustainability which are based on assumptions more concerned with the "human" aspect and neglect or minimize the enforcement of the "ecological principles" set forth in the very own definition of sustainable construction. Neither do they seem to incorporate the severity of the environmental impacts caused by the human specie and already described in Sect. 1.1. Although the construction sector is not confined to the building sector (residential or otherwise), also covering the civil engineering works related to infrastructures, it is symptomatic that the emphasis on sustainable construction has been placed on the energy efficiency of the building stock. One reason for this, perhaps the most important, has to do with the high cost of buildings on operational energy consumption. This is not the case for other civil engineering works, in which the environmental performance is crucial only at the time of the execution and conditioned by the construction materials used in it. However, as we will show in Chap. 3, as the buildings become increasingly energy efficient, the role of energy related to construction and building materials (embodied energy) acquires increasing importance, justifying an increased attention.

1.4 Eco-Efficient Construction and Building Materials

The concept of eco-efficiency was first introduced in 1991 by the World Business Council for Sustainable Development-WBCSD (2000) and includes "*the development of products and services at competitive prices that meet the needs of humankind with quality of life, while progressively reducing their environmental impact and consumption of raw materials throughout their life cycle, to a level compatible with the capacity of the planet*". This concept means producing more products with fewer resources and less waste (Bidoki et al. 2006). Therefore, the eco-efficient construction and building materials are those that present a less environmental impact. It is then necessary to assess all environmental impacts caused by a given material since the beginning of the extraction of raw materials (cradle) to the end of its service life (grave).

One of the most important environmental issues related to the production of construction and building materials, are the environmental impacts caused by the extraction of raw materials (Barnett and Morse 1963; Suslick and Machado 2009). Other impacts will be analyzed in the subsequent chapters of this book. For some decades several authors and organizations have warned about the depletion of raw materials, and some even pointed to early doomsday scenarios unproven so far, the most famous of which has been the report "*The Limits to Growth*" (Meadows et al. 1972) produced by the Massachusetts Institute of Technology (MIT) to the Club of Rome, a think tank founded in 1968 by scientists, politicians and entrepreneurs. While it is important to recognize that there are exhaustible non-renewable raw materials, it is not clear that its duration is that presented in Table 1.4. The problem of consumption scarcity would be more serious if the high consumption patterns of the developed countries expanded throughout the world population. The real environmental threat associated with the production of construction and building materials, is not so much the depletion of non-renewable raw materials, but instead the environmental impacts caused by its extraction (Meadows et al. 2004). Including the destruction of the biodiversity of mining areas, the threats associated to mining wastes (Table 1.5) and the possible environmental accidents caused by them.

In 2000, the mining activity worldwide generated 6000 million tons of mine wastes to produce just 900 million tons of raw materials (Whitmore 2006). This means an average use of only 0.15%, resulting in vast quantities of waste, whose

Table 1.4 Non-renewable raw materials (Berge 2009)

Raw material	Reserves (years)	Annual growth consumption between 1999 and 2006 (%)
Agreggates	Very large	–
Arsenic	20	6
Bauxite	141	6
Bentonite	Large	–
Boric salts	35	1
Brom	Large	–
Cadmium	26	1
Clay	Very large	–
Chrome	25	8
Coal	150	4,5
Cobalt	121	15
Copper	31	3
Crude oil	41	1,4
Diatomite	Large	–
Earth for compressing	Very large	–
Feldspar	Large	–
Gold	17	1
Iron	95	10
Kaolin	Large	–
Lime	Very large	–
Lead	20	1,5
Magnesium	Large	–
Manganese	40	9
Natural gas	63	3
Nikel	41	5
Perlite	Large	–
Phosphate	124	–
Potash	Large	–
Pumice	Large	–
Quartz	Large	–
Silver	14	3
Silica	Large	–
Sodium carbonate	Large	–
Sodium chloride	Very large	–
Sulphur	21	1
Tin	22	4
Titanium	122	5
Vermiculite	Large	–
Zinc	22	4,5

disposal represents an environmental risk in terms of biodiversity conservation, air pollution and pollution of water reserves. In 2003, the US mining activities were responsible for releasing 3 billion pounds of toxic substances (Royte 2005). As a

Table 1.5 Amount of rock excavated and percentage of raw material extracted (Gardner and Sampat 1998)

Raw material	Excavated rock (MT)	Extracted raw material (%)
Iron	25503	40
Copper	11026	1
Zinc	1267	0.05
Aluminum	869	30
Lead	1077	2.5
Tin	195	1
Nikel	387	2.5
Tungsten	125	0.25
Manganese	745	30

Fig. 1.5 Mine of Aznalcollar—view of the landfill rupture

result, since the 70s there were 30 serious environmental accidents in mines, and 5 of them occurred in Europe. The following is a description of three mine accidents in Europe:

In July 1985, two embankments that contained mine sludge, located in the valley of Stava in Italy broke up and a torrent of fines and sands razed 68 buildings and killed 268 people (Alexander 1986). In April 1998, the toxic landfill sludge of the Aznalcollar Mine in Spain (Fig. 1.5), broke up, and almost five million tons of mine sludge were released into the Agrio river.

The mine sludge spread over a length of 40 km contaminating 2,650 hectares of the Donana National Park, a Unesco world heritage (Grimalt and Macpherson 1999). Investigations carried out on plant species revealed high concentrations of heavy metals (Gómez-Parra et al. 2000), which in some cases reached values 100 times higher than the current levels. The same investigations also showed that many animal species had high levels of heavy metals in their bodies. On January 30, 2000, a landfill sludge in Baia Mare mine, Romania broke up and 100,000 m^3

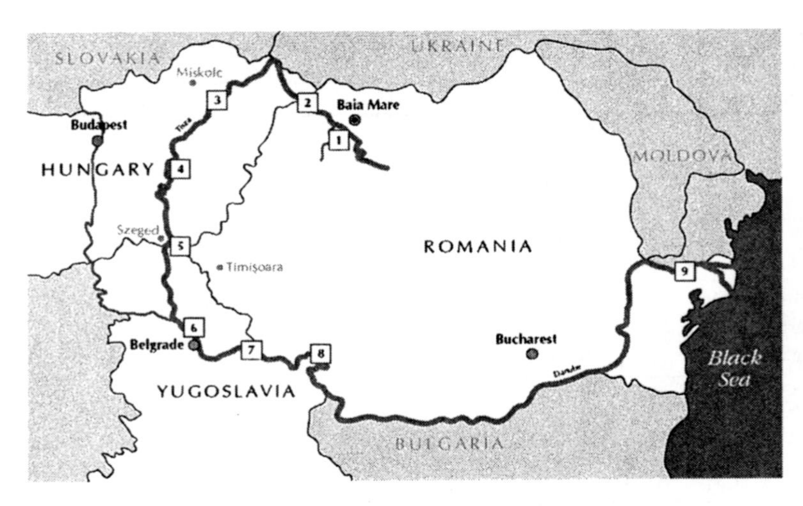

Fig. 1.6 Mina of Baia Mare—evolution of the sludge spreading from Romania until the Black Sea (Csagoly 2000)

of sludge contaminated with cyanide, arsenic and other metals were released into the river Tisza (Fig. 1.6 n-1). On February 1 the contamination was on the borders of Hungary, on February 11th reached Yugoslavia and by the end of February it reached the Black Sea (Fig. 1.6 n-9). High concentrations of cyanide were detected up to 2000 km away from the place of the sludge discharge (Michnea and Gherhes 2001; Kraft et al. 2006).

The mine accidents, particularly the two that took place in 1998 and 2000 respectively in Spain (Los Frailes) and in Romania (Baia Mare), show that within a short period the environmental impacts associated with mining operations, represent a threat as serious as climate change (BRGM 2001; Puura et al. 2002). Apart from the need to minimize the extraction of non-renewable raw materials, other issues must be considered in the context of eco-efficient construction and building materials. The choice of these materials must focus on:

- non toxic materials;
- materials with low embodied energy;
- recyclable materials;
- materials containing wastes from other industries;
- materials obtained from renewable resources;
- materials responsible for low GHGs;
- materials with high durability;
- materials with self-cleaning ability and capable of reducing air pollution.

A summary of the contents of each of the remaining ten chapters that constitute the present book is presented below.

Chapter 2 presents several construction and building materials associated with some level of toxicity, whether in terms of the contamination of indoor air, the realease of toxic fumes during fires or even by contaminating drinking water. This chapter also contains considerations about the importance of leaching tests to assess the potential hazard of construction and building materials containing other wastes. Chapter 3 deals with the importance of reducing energy consumption as a crucial aspect of sustainable construction. It covers the embodied energy and also the materials responsible for the reduction of energy consumption. Namely current thermal insulation materials, insulation materials based on vegetable fibres, high performance insulation materials and also phase change materials. Chapter 4 addresses the subject of construction and demolition wastes. It covers the waste volume estimation, the deconstruction principles, the sorting and recycling procedures. The recycling of gypsum wastes, asbestos and concrete receive close attention. Chapter 5 presents several contributes in order to increase the eco-efficiency of concrete, the most used construction material on Earth. By replacing natural aggregates with industrial wastes (construction and demolition wastes, recycled tyres and poly-ethylene terephthalate wastes) or even by replacing cement with industrial by-products (fly ash, silica fume, rice husk ash, sewage sludge ash, ceramic wastes, recycled glass). The chapter also describes the characteristics of organic polymer concrete, concrete with sensing capability, and also new family binders alternative to Portland cement. Chapter 6 compares the contribution of ceramic bricks and concrete blocks towards the eco-efficiency of masonry. The incorporation of wastes from other industries as well as shape optimization are considered. Chapter 7 covers the use of vegetable fibres as reinforcement in cement-based composites. It includes fibre characteristics, properties and the description of the treatments that improve their performance; it also includes the properties and durability performance of cementitious materials reinforced with vegetable fibres. Chapter 8 is related to earth construction; it describes the construction techniques, the soil selection and the soil stabilization procedures and also covers durability-related issues. Chapter 9 deals with the pathology and durability of concrete. It also covers the pathology of gypsum and lime-based renders, the procedures to evaluate the composition of renders of ancient buildings and the criteria used in their rehabilitation. Chapter 10 examines the role of nanotechnology to increase the eco-efficiency of construction and building materials. It covers the nanoscale analysis of Portland cement hydration products, the use of nanoparticles to increase the strength and durability of cimentitious composites and also the photocatalytic capacity of nanomaterials. Chapter 11 deals with LCA of construction and building materials. It covers eco-labeling and environmental product declarations (EPD's). Some practical cases are presented in this chapter.

1.5 Conclusions

The Earth faces very serious environmental threats that require immediate action.

One must realize that even if all the carbon dioxide emissions suddenly ceased, the amount already in the atmosphere would remain there for the next 100 years

(Clayton 2001). Unfortunately worldwide consensus on this matter that would translate into meaningful action has not yet been reached. This is because different countries have different levels of development and those in lower positions believe that any environmental constraints imposed on them will prevent their economic ascent. Since the Earth does not have an unlimited capacity in terms of continuing to support the current level of devastation, the more developed countries must adopt very high environmental standards. The construction industry is one of the sectors that must reduce its environmental impact. Several tools have been developed for the assessment of the environmental impacts of buildings; however, it seems unlikely that it is possible to accommodate the 8,000 million people expected for the year 2050, in buildings with the highest environmental rating under these tools without intolerable environmental consequences. As for construction and building materials, their highest environmental impact is not so much related to their scarcity but to the extraction of raw materials needed for their production.

References

Alexander D (1986) Northern Italian dam failure and mud flow, July 1985. Disasters 10:3–7. doi: 10.1111/j.1467-7717.1986.tb00560.x

Allan R, Soden B (2008) Atmospheric warming and the amplification of precipitation extremes. Science 321:1481–1484. doi:10.1126/science.1160787

Anthony K, Kline D, Diaz-Pulido G, Dove S, Hoegh-Guldberg O (2008) Ocean acidification causes bleaching and productivity loss in coral reef builders. Proc National Acad Sci USA 105:17442–17446

Balmford A, Bruner A, Cooper P, Costanza R, Farber S, Green RE, Jenkins M, Jefferiss P, Jessamy V, Madden J, Munro K, Myers N, Naeem S, Paavola J, Rayment M, Rosendo S, Roughgarden J, Trumper K, Turner K (2002) Economic reasons for conserving wild nature. Science 297:950–953. doi:10.1126/science.1073947

Barnett H, Morse C (1963) In: Smith VK (ed), Scarcity and growth. Johns Hopkins University Press, Baltimore

Berge B (2009) The ecology of building materials, 2nd edn. Architectural Press, Oxford

Bidoki S, Wittlinger R, Alamdar A, Burger J (2006) Eco-efficiency analysis of textile coating agents. J Iran Chem Soc 3:351–359. http://www.ics-ir.org/jics/archive/v3/4/article/pdf/JICS-3-4-Article-8.pdf. Accessed 2 July 2011

Bourne J (2008) Temperature has already changed. National Geographic Magazine, Washington

BRGM (2001) Management of mining, quarrying and ore-processing waste in the european union. European Commission, DG environment, 50319-FR

Bribian I, Uson A, Scarpellini S (2009) Life cycle assessment in buildings: State-of-the-art and simplified LCA methodology as a complement for building certification. Build Environ 44:2510–2520. doi:10.1016/j.buildenv.2009.05.001

Brittain C, Vighi M, Bonmmarco R, Settele J, Potts S (2010) Impacts of a pesticide on pollinator species richness at different spatial scales. Basic Appl Ecol 11:106–115. doi: 10.1016/j.baae.2009.11.007

Broecker W, Kunzig R (2008) Fixing climate: what past climate changes reveal about current threat and how to counter it. Hill and Wang, New York

Bruntland G (1987) Our common future. Report of the World Commission on Environment and Development. Oxford University Press, Oxford

Chapman L (2007) Transport and climate change: a review. J Transp Geography 15:354–367. doi:10.1016/j.jtrangeo.2006.11.008

Choi J, Pattent B (2001) Sustainable development: lessons from the paradox of enrichment. Ecosystem Health 7:163–175. doi:10.1046/j.1526-0992.2001.01024.x

Clayton R (2001) Editorial: Is sustainable development an oxymoron? Trans I Chem E 79 (Part B):327–328

Copenhagen Climate Summit (2009) United Nations Framework Convention on Climate Change

Costanza R, D'arge R, De Groot R, Farber S, Grasso M, Hannon B, Limburg K, Naeem S, O'Neill R, Paruelo J, Raskin R, Sutton P, Van Den Belt M (1998) The value of the world's ecosystem services and natural capital. Nature 387:253–260. doi:10.1038/387253a0

Csagoly P (2000) The cyanide spill at baia mare, Romania. Before, during and after. Regional Environmental Center for Central and Eastern Europe

Derraik J (2002) The pollution of the marine environment by plastic debris: a review. Mar Pollut Bull 44:842–852. doi:10.1016/S0025-326X(02)00220-5

Destatte P (2010) Foresight: a major tool in tackling sustainable development. Technol Forecasting Soc Change 77:1575–1587. doi:10.1016/j.techfore.2010.07.005

Dietz T, Rosa E, York R (2009) Environmental efficient well-being: Rethinking sustainability as the relationship between the human well-being and Environmental impacts. Hum Ecol Rev 16:114–123

Dobbs R (2010) Prime numbers: megacities. foreign policy. http://www.foreignpolicy.com/articles/2010/08/16/prime_numbers_megacities. Accessed 10 October 2010

Fenner R, Ryce T (2008a) A comparative analysis of two building rating systems. Part 1: Evaluation. Proc Inst Civ Eng: Eng Sustainability 161:55–64

Fenner R, Ryce T (2008b) A comparative analysis of two building rating systems. Part 2: A case study. Proc Inst Civ Eng: Eng Sustainability 161:65–70

Gardner G, Sampat P (1998) Mind over matter: recasting the role of materials in our lives. Worldwatch Institute, Washington

Genersch E (2010) Honey bee pathology: Current threats to honey bees and beekeeping. Appl Microb Biotech 87:87–97. doi:10.1007/s00253-010-2573-8

Goldenberg J, Prado L (2010) The "decarbonization" of the world's energy matrix. Energy 38:3274–3276. doi:10.1016/j.enpol.2010.03.040

Gómez-Parra A, Forja J, Del Valls T, Sáenz I, Riba I (2000) Early contamination by heavy metals of the Guadalquivir estuary after the Aznalcóllar mining spill (SW Spain). Mar Pollut Bull 40:1115–1123. doi:10.1016/S0025-326X(00)00065-5

Gore A (2009) Our choice. A plan to solve the climatic crisis. Rodale Press, Emmaus

Grimalt J, Macpherson E (1999) The environmental impact of the mine tailing accident in Aznalcóllar (S.W. Spain). Science Total Environ 242 (Special issue)

IPCC (2007) Intergovernmental panel on climate change, climate change 4th assesment report

Kibert C (1994) Establishing the principles and model for sustainable construction. In Proceedings of the 1st international conference of CIB TG 16: 1–9, Tampa

Kibert C (2008) Sustainable construction: green building design and delivery, 2nd edn. Wiley, Hoboken

Kraft C, Tumpling J, Zachmznn D (2006) The effects of mining in Northern Romania on the heavy metal distribution in sediments of the rivers Szamos and Tisza (Hungary). Acta Hydrochimica Hydrobiologica 34:257–264. doi:10.1002/aheh.200400622

Lee W, Burnett J (2008) Benchmarking energy use assessment of HK-BEAM, BREEAM and LEED. Buil Environ 43:1882–1891. doi:10.1016/j.buildenv.2007.11.007

Liu S, Fu C, Shiu C, Chen J, Wu F (2009) Temperature dependence of global precipitation extremes. Geophys Res Lett 36, no L17702. doi: 10.1029/2009GL040218

Mateus R (2009) Assessment of construction sustainability. Proposals for the development of more sustainable buildings. PhD Thesis, School of Engineering, University of Minho, Portugal

Meadows D, Randers J, Meadows D (1972) The limits to growth: a report for the Club of Rome's project on the predicament of mankind. Universe Books, New York

Meadows D, Randers J, Meadows D (2004) The limits to growth: a 30-year update. Chelsea Green Publishing, Vermont

Meehl GA, Stocker TF, Collins WD, Friedlingstein P, Gaye A, Gregory J, Kitoh A, Knutti R, Murphy J, Noda A, Raper SC, Watterson IG, Weaver A, Zhao Z (2007) Global Climate Projections. In: Solomon S, Qin D, Manning M, Chen Z, Marquis M, Averyt K, Tignor M, Miller H (eds) Climate Change 2007: The Physical Science Basis. Contribution of Working Group I to the Fourth Assessment Report of the Intergovernmental Panel on Climate Change. Cambridge University Press, Cambridge

Michnea A, Gherhes I (2001) Impact of metals on the environment due to technical accident at Aurul Baia Mare, Romania. Int J Occup Med Environ Health 14:255–259

Micklin P (2007) The Aral sea disaster. Annu Rev Earth Planet Sci 35:47–72. doi: 10.1146/annurev.earth.35.031306.140120

Millenium (2005) Ecosystem assessment synthesis report. In: Whyte A, Sarukhan J (eds) . World Resources Institute, Washington

Moore C (2008) Synthetic polymers in the marine environment: a rapidly increasing, long-term threat. Environ Res 108:131–139. doi:10.1016/j.envres.2008.07.025

Moore C, Moore S, Leecaster M, Weisberg S (2001) A comparison of plastic and plankton in the North Pacific central gyre. Mar Pollut Bull 42:1297–1300. doi:10.1016/S0025-326X(01)00114-X

Pearce A, Walrath L (2008) Definitions of sustainability from the literature. Sustainable facilities and infrastructure. Georgia Institute of Technology, Atlanta

Puura E, Marmo L, Marco A (2002) Mine and quarry waste—the burden from the past. Directorate General Environment of the European Commission, Lake Orta

Rees W, Wackernagel M (1966) Urban ecological footprints: why cities cannot be sustainable—And why they are a key to sustainability. Environ Impact Assess Rev 16:223–248. doi: 10.1007/978-0-387-73412-5_35

Ring I, Hansjurgens B, Elmqvist T, Wittmer H, Sukhdev P (2010) Challenges in framing the economics of ecosystems and biodiversity: the TEEB initiative. Curr Opin Environ Sustainability 2:15–26. doi:10.1016/j.cosust.2010.03.005

Robinson J (2004) Squaring the circle? Some thoughts on the idea of sustainable development. Ecol Econ 48:369–384. doi:10.1016/j.ecolecon.2003.10.017

Royte E (2005) Garbage land: on the secret trail of trash. Little, Brown and Company, UK

Saunders M, Lea A (2008) Large contribution of sea surface warming to recent increase in Atlantic hurricane activity. Nature 451:557–560. doi:10.1038/nature06422

Schellnhuber H (2008) Global warming. Stop worrying, start panicking? Proceedings of the National Academy of Sciences of the United States of America 105:14239–14240. doi: 10.1073/pnas.0803838105

Shealy M, Dorian J (2009) Growing Chinese coal use: dramatic resource and environmental implications. Energy Policy 38:2116–2122. doi:10.1016/j.enpol.2009.06.051

Spiertz J (2010) Nitrogen, sustainable agriculture and food security. A review. Agronomy for Sustainable Development 30:43–55. doi:10.1051/agro:2008064

Stern N (2006) Stern review on economics of climate change. Cambridge University Press, Cambridge

Suslick S, Machado I (2009) Non-renewable resources. in earth system: history and natural variability. In: Cilek V, Smith R (eds) Encyclopedia of life support systems. EOLSS Publishers, Oxford

UN (2008) Revision of the world population prospects. UN Publications, New York

UNEP (2006) Climate change scenarios for desert areas. In UNEP/GRID-Arendal Maps and Graphics Library. UN Publications, New York

UNFCCC (1997) Kyoto Protocol to the United Nations framework convention on climate change, United Nations convention on climate change, FCC/CP/L.7/Add1, Kyoto. UN Publications, New York

Vellinga M, Wood R (2002) Global climatic impacts of a collapse of the Atlantic termohaline circulation. Clim Change 54:251–267. doi:10.1023/A:1016168827653

Vellinga M, Wood R (2008) Impacts of thermohaline circulation shutdown in the twenty-first century. Climatic Change 91:43–63. doi:10.1007/s10584-006-9146-y

Wang T, Watson J (2010) Scenario analysis of China's emissions pathways in the 21st century for low carbon transition. Energy Policy 38:3537–3546. doi:10.1016/j.enpol.2010.02.031

Whitmore A (2006) The emperors new clothes: Sustainable mining. J Clean Prod 14:309–314

Wiedmann T, Barrett J (2010) A Review of the ecological footprint indicator—perceptions and methods. Sustainability 2:1645–1693. doi:10.3390/su2061645

WWF (2008) Zoological Society of London; Global Footprint Network; Twente Water Centre. Living Planet Report 2008; World-Wide Fund for Nature International (WWF): Gland

Zolina O, Simmer C, Gulev S, Kollet S (2010) Changing structure of European precipitation: longer wet periods leading to more abundant rainfalls. Geophys Res Lett 37, no L06704. doi: 10.1029/2010GL042468

Chapter 2
Toxicity of Construction and Building Materials

2.1 General

While our ancestors lived in buildings made of raw materials, nowadays residential buildings contain a high amount of chemicals and heavy metals, that either contaminate indoor air or pollute tap water, thus causing several health-related problems such as: asthma; itchiness; burning eyes, skin irritations or rashes, nose and throat irritation; nausea; headaches; dizziness; fatigue; reproductive impairment; disruption of the endocrine system; cancer; impaired child development and birth defects; immune system suppression and cancer. Besides the toxicity of building materials indoors one must also keep in mind the toxicity potential during the production of chemicals. Since 1930 more than 100,000 new chemical compounds have been developed and insufficient information exists for health assessments of 95% of chemicals in the environment (World Watch Institute 1998). Recall for instance the Bhopal disaster that occured in India in 1984, when a cloud of methyl isocyanate caused almost 15,000 deaths and health problems in almost 200,000 human beings (Varma and Mulay 2006; Satyanand 2008). During the production of chemical materials hazardous wastes are generated and those impacts should be related to the building materials containing these chemicals. The most common toxic chemicals are the following: Dioxins and furans—Dioxins and furans are chemical wastes generated in the industrial process evolving chlorine as in PVC production. Dioxins and furans are extremely toxic and also bio-cumulative (Koopman-Esseboom et al. 1996; Paauw et al. 1996; IARC 1997; Lanting et al. 1998). This has hazardous effects on biodiversity by contaminating the entire food chain (Oppenhuizen and Sijm 1990; Tillitt et al. 1993). Furthermore, chemical analyses carried out in dolphins in the North pacific ocean reveal dioxin and furan concentrations between 13 million and 37 million times higher than plain sea water concentrations (Thorton 2000). Several scientist groups already suggest that chlorine industrial-based products should be prohibited (Flores et al. 2004). Phthalates—chemical compounds caused by phthalic acid. Phthalates are used to soften plastic materials. Several studies show that phthalates are very toxic

F. P. Torgal and S. Jalali, *Eco-efficient Construction and Building Materials*,
DOI: 10.1007/978-0-85729-892-8_2,

Table 2.1 Cancer agents detected in paints (IARC 1995; UNCHS 1997)

Cancer agent	Likely source
Chromates	Primers, paints
Cadmium	Pigments
Benzene	Solvents
Methylene chloride	Paint strippers
Styrene	Organic solvents
Nikel compounds	Pigments
Dichlorobenzidine	Pigments
Lead	Primers, dryers, pigments
Antimony oxide	Some pigments
Nitropropane	Organic solvent
Tetrachloroethylene	Organic solvent

to human health (Lovekamp-swan and Davis 2003; Hauser and Calafat 2005; Heudorf et al. 2007; Wolff et al. 2008; Swan 2008; Meeker et al. 2009).Volatile Organic Compounds (VOCs)—Atmospheric pollutants release from building materials which contain organic solvents like paints and varnishes. The reduction of indoor ventilation to minimize energy consumption (as it often happens in many countries) contributes to increase the effects of VOCs in human health (Sterling 1985; Samfield 1992; Hansen and Burroughs 1999).

2.2 Paints, Varnishes and Wood Impregnating Agents

Several authors confirm the release of VOCs from paints and varnishes (Kostianien 1995; Kwok et al. 2003). More recently Salasar (2007) studied VOCs emissions in solvent and water-based paints stating that the former are responsible for VOCs emissions 520 higher than the latter. Painting can also be a source of several cancer agents (Table 2.1).

Although wood is an excellent example of a sustainable building material it has low resistance to biologic degradation (fungal and insect attack) (Morrell 2002). Until very recently wood preservation meant the use of impregnating agents (insecticides or fungicides) like creosote or others based on salt impregnation like copper, chrome and arsenic (CCA). However, these salts are highly toxic and also bio-cumulative. When they come in contact with rain water the majority of these salts is leached away contaminating the environment. Since January 1, 2004 the EPA-USA forbid the use of CCA for wood preservation (Edlich et al. 2005). Regarding creosote, it contains potential cancer agents (ATSDR 2002; Smith 2008) therefore, since the 2001 Directive 2001/90/EC, the process of banning the use of creosote for wood treatment purposes was initiated. Some studies mentioned that wood used in railway cross ties has a high content of creosote (Thierfelder and Sandstrom 2008), so they must be seen as hazardous wastes, meaning that they must be properly immobilized and can no longer be reused (Pruszinski 1999). The same should apply to all the creosote-treated wood which in the near future ends up in construction and demolition wastes.

2.3 Plastics and Synthetic Adhesives

Plastics are used in the production of several building materials and come from the extraction of non-renewable resources by the oil industry. Plastics are generated by a polymerization of basic molecules (monomers), leading to long chain monomers, which fall into two main categories, thermoplastics and thermosetting plastics. The former are the plastics supplied ready to use that can be softened by temperature and become rigid on cooling such as PVC, polyethylene, polypropylene or polystyrene. The thermosetting plastics are supplied as fluid products that acquire their final and irreversible shape by chemical reaction (two part epoxi); or when submitted to a temperature above 200°C. This group includes polyurethane, melamine, styrene butadiene rubber or epoxy and other synthetic adhesives. Almost all plastics contain various additives such as plasticizers, softners, heat stabilizers, UV stabilizers, dyes, smoke reducers and others that imply the use of phthalates and heavy metals. Polystyrene is obtained from the polymerization of styrene, and its applications relates to thermal insulation products, obtained by expansion (EPS) or extrusion (XPS). This material contains anti-oxidant additives and ignition retardants and during its production benzene and chlorofluorocarbons are generated. Polyethylene is obtained from the polymerization of ethylene, containing 0.5% of additives such as phenol based anti-oxidants, UV stabilizers and dyes, including aluminum, magnesium hydroxide, chloroparaffin as ignition retardants. Polypropylene is obtained from the polymerization of propylene with additives similar to those used in polyethylene (Berge 2009). Polyurethane is obtained from isocyanates, known worldwide for its tragic association with the Bhopal disaster.This substance is highly toxic (Marczynski et al. 1992; Baur et al. 1994) and there are multiple records of serious health problems in workers using polyurethane (Littorin et al. 1994; Skarping et al. 1996). Chester et al. (2005) even reported the death of a worker due to the simple application of polyurethane. The production of polyurethane also involves the production of toxic substances such as phenol and chlorofluorocarbons. Polyvinyl chloride (PVC) is a thermoplastic polymer, obtained from the polymerization of vinyl chloride monomer, which in turn is obtained from petroleum and chlorine. Worldwide consumption of PVC is approximately 30 million tons annually and it is used mostly in pipe production (Rahman 2007). Despite being the third most produced plastic, after polyethylene and polypropylene, PVC plastic is the largest producer of organochlorine substances (Thorton 2000, 2002). Furthermore, the production of PVC involves the use of tin-based stabilizers (organostannic compounds) in order to prevent this material to degrade under the effects of temperature (between 150°C and 200°C). This compound has a high eco-toxicity level and may have a harmful effect on the environment (Hoch 2001) even in terms of contamination of water supply (Wu et al. 1989; Sadiki and Williams 1996, 1999; Forsyth and Jay 1997; Sadiki et al. 1996; Stern and Lagos 2008; Fristachi et al. 2009). Beyond the environmental impacts associated with the production of plastic materials, it should also be taken into account that these materials are not biodegradable and their end of life treatment involves GHGs emissions. Synthetic adhesives are used in the construction industry for several purposes, which include bonding laminated wood, waterproofing materials or even the rehabilitation of concrete structures. In terms of their composition they can be based on epoxy resins, melamine-urea-formaldehyde, phenol or organic solvents. The epoxy resins are toxic materials and the workers exposed to them can show eczemas and dermatitis. These materials are also responsible for trigging the appearance of allergies and even cancer (Peltonen et al. 1986; Tsai 2006). The melamine-urea-formaldehyde compounds are equally toxic, and some authors argue that they have carcinogenic potential (Vale and Rycroft 1988; Wilbur et al. 1999; Zhang et al. 2008). Adhesives based on organic solvents are also hazardous (Heuser et al. 2005).

2.4 Materials That Release Toxic Fumes During Fire

Another case of building materials toxicity is related to materials that release toxic fumes during fires. Some studies show that the majority of deaths during fires is due to the inhalation of toxic fumes and that such deaths are increasing since the 1980s, maybe due to the fact that the amount of combustible materials inside households has increased in the last three decades (Gann et al. 1994; Hall and Harwood 1995; Wu 2001; Levin and Kuligowski 2005). Liang and Ho (2007) studied the toxicity during fires of several insulation materials concluding that polyethylene foam and polyurethane foam have a toxicity index (TI) higher than 10, thus meaning extremely high toxicity (Fig. 2.1) levels.

The TI is obtained from the emissions of 14 different combustion gases reaching deadly concentration after 30 minutes exposure. These authors recommend that polyethylene and polyurethane foams should not be used unless covered by incombustible materials. Other authors Doroudiani and Omidian (2010) say that polystyrene decorative mouldings should be avoided because polystyrene is a very combustible material that releases toxic fumes during fires. They also say that new polystyrene with flame-retardant properties is now under production, but it also releases other toxic substances.

2.5 Radioactive Materials

The use of waste materials with some kind of radiological contamination is known to be a matter of concern to public health because exposure during a long period even to low doses can lead to cancer formation (ICRP 1990). In general, building

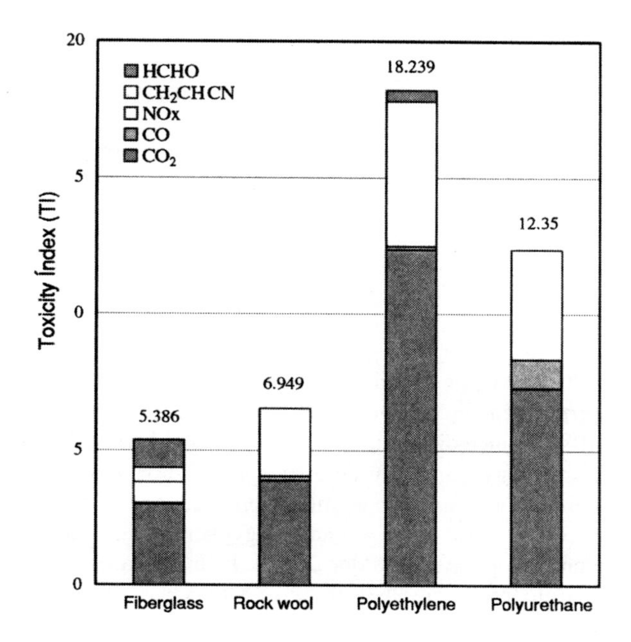

Fig. 2.1 Average value of toxicity index (Liang and Ho 2007)

Table 2.2 Typical and maximum activity concentrations in common building materials and industrial by-products used for building materials in Europe (Kovler et al. 2002; Kovler 2009)

Material	Typical activity concentration (Bq/kg)			Maximum activity concentration (Bq/kg)		
	^{226}Ra	^{232}Th	^{40}K	^{226}Ra	^{232}Th	^{40}K
Construction materials						
Concrete	40	30	400	240	190	1600
Lightweight concrete	60	40	430	2600	190	1600
Ceramic bricks	50	50	670	200	200	2000
Concrete blocks	10	10	330	25	30	700
Natural stone	60	60	640	500	310	4000
Natural gypsum	10	10	80	70	100	200
Industrial by-products						
Phosphogypsum	390	20	60	1100	160	300
Blast furnace slag	270	70	240	2100	340	1000
Coal fly ash	180	100	650	1100	300	1500

materials do not show dangerous radioactivity levels (Papaefthmiou and Gouseti 2008), but the same cannot be said about some industrial by-products used for concrete production such as phosphogypsum, some blast furnace slags and some fly ashes (Table 2.2).

Phosphogypsums contains heavy metals and radioactive elements such as radio (^{226}Ra), lead (^{210}Pb) e uranium (^{238}U, ^{234}U) that come from phosphate rocks (Rihanek 1971). The use of phosphogypsum with a concentration level of

Table 2.3 Radom concentration (Bq/m^3) due to radon exhalation from floor material according to the ventilation rate (Chen et al. 2010)

Radon exhalation rate (Bq/m^{2}d)	Air changes per hour (ACH)				
	3	1	0.3	0.15	0
5	0.03	0.09	0.3	0.6	5
10	0.06	0.2	0.6	1.2	25
50	0.3	0.9	3.0	5.9	123
100	0.6	1.8	6.0	12	246
300	1.8	5.5	18	35	737

370 Bq/kg (in which 1 Bq corresponds to 1 nuclear disintegration per second), is prohibited since 1992 (EPA 1992). The Euratom threshold is 500 Bq/kg (Euratom 1996). Since different phosphate rocks possess different radioactivity levels, not all phosphogypsums can be considered radioactive (Canut 2006). Another important case of radioactive contamination is related to radon exhalation, a radioactive gas found in some types of phosphogypsum which can be toxic for indoor air with low ventilation rates (Kovler 2009). In most cases radon comes from the ground in granite areas but its source can be from granite floor materials thus polluting indoor air. Chen et al. (2010) analyzed 33 different types of granites and mentioned that only two of them had exhalation rates above 200 (Bq/m^{2}d). These findings were confirmed by Pavlidou et al. (2006). Chen et al. (2010) studied the combined influence of indoor air ventilation rate and granite exhalation rates serving as floor materials, concluding that the highest exhalation rate granite from floor materials in a place with a low ventilation rate (ACH $= 0.3$) contributes only with 18 Bq/m^3 to the total concentration (Table 2.3).

For ACH levels near zero, high exhalation rate granites can be effectively responsible for toxic radioactive concentrations. Other cases of radioactive building materials can be found in the literature. In Sweden, 300,000 residential buildings were made from concrete based on aggregates from a uranium mine (alum shale). Recent studies reveal that infants and children are more prone to develop leukemia-related diseases (Axelson et al. 2002).

2.6 Asbestos-Based Materials

Asbestos covers several mineral fibres with 5 μm length and 3 μm in diameter such as: chrysolite, crocidolite, amosite, anthhrophyllite, tremolite and actinolite. It was not until the 1960s that a relationship between asbestos exposure and several professional diseases was established by scientific evidence. By that time only some mineral fibres (crocidolite-blue asbestos and amosite-brown asbestos) were thought of as toxic and responsible for pleural mesothelioma from which most patients died, just 12 months after being diagnosed (Bianchi et al. 1997; Jarvholm et al. 1999; Azuma et al. 2009). Chrysolite-white asbestos was not included because it was considered to have a low toxicity risk and that is why asbestos continues to be produced. Only in the 1980s with the Directive 83/477/CEE did the asbestos problem start to be taken more seriously. Since the Directive 91/382/EEC was enforced even stricter caution about asbestos was imposed and finally Directive 2003/18/EC prohibited the production of asbestos-based products. This Directive defines a threshold risk (VLE) when asbestos fibres concentration is higher than 0.1 fibres/cm^3. In the meantime, scientific evidence proved that all minerals fibres present cancer risks such as asbestosis (lung damage due to acid formation in a body's attempt to dissolve the asbestos fibres) (Akira 2010) or even lung cancer or other types of cancer (Ladou 2004; Silverstein et al. 2009; Antonescu-Turcu and Schapira 2010). Although some may think that asbestos is no longer a problem let's not forget about the vast number of fibres-cement materials asbestos based that are still installed. One may argue that cement materials containing asbestos has low toxicity risk but it is also true that cement will lose its binder capacity under environmental erosion and some cracking accident could take place releasing asbestos fibres. Therefore, it is not possible to say that people working (or living) under fibre-cement asbestos roofing sheets are not submitted to a fibre concentration higher than the VLE threshold or if people submitted to a fibre concentration below VLE will not develop cancer after a long time of exposure. It is noteworthy that according to World Health Organization (WHO) there is no such thing as a safe asbestos threshold.

2.7 Nanoparticles

Although the use of nanoparticles is very recent, it has already raised issues concerning its potential toxicity. Some investigations show that nanoparticles can cause symptoms like those caused by asbestos fibres. Grassian et al. (2007) studied the effects related to the inhalation of TiO_2 particles with primary particle sizes between 2 and 5 nm, reporting lung inflammation for a concentration of 8.8 mg/m^3. These symptoms have been confirmed by other authors (Yu et al. 2008; Liu et al. 2008; Poland et al. 2008; Donaldson and Poland 2009; Pacurari et al. 2010). Hallock et al. (2008) recommending that the use of nanoparticles should be made with the same care already used in Universities for materials of unknown toxicity, i.e., by using air extraction devices to prevent inhalation and gloves to prevent dermal contact. Singh et al. (2009) mentioned the possibility of DNA damage resulting in later cancer development. Some authors (Dhawan et al. 2009) believe that the nanotoxicity risk depends on the nanoparticles type, concentration volume and superficial characteristics. Other authors (Tyshenko and Krewski 2008) suggest that several categories and new parameters must be formulated to better analyze this subject. Some questions that deserve further investigation are presented elsewhere (Walker and Bucher 2009; Hirano 2009). Bystrzejewska-Piotrowska et al. (2009) have recently carried out an extended literature review on this subject. These authors mentioned that Environment Protection Agency has considered that carbon nanotubes are a new form of carbon that must be treated under the toxic products act. These authors also mentioned that nanoparticles may be responsible for a new kind of problem, the appearance of nanowastes. They suggest that products containing nanoparticles should be labeled in order to facilitate future separation and recycling procedures.

2.8 Lead Plumbing

Due to its low corrosion characteristics lead was used as water pipe material at least since the Roman Empire (Hodge 1981; Dutrizac et al. 1982; Nriagu 1983). Several authors mentioned that lead plumbing is responsible for health problems, because a film of corrosion products is formed at the pipe surface that eventually will be leached away, thus contaminating water (Zietz et al. 2009). This contamination is particularly toxic to infants and children causing behavior problems and intellectual impairment (Pocock et al. 1994; Wilhelm and Dieter 2003; Canfield et al. 2003). Troesken (2006) refers several cases of lead poisoning due to lead plumbing in the last two centuries. A problem that is as big as the Chernobyl or Bhopal disasters. This author states that only in the US thousands of children have died due to lead poisoning and as much as that amount suffered from intellectual impairment. A blood lead content higher than 10 μg/dl is considered to be the threshold of lead poisoning (Labat et al. 2006; Tararbit et al. 2009), it is also associated with cardiovascular related deaths and cancer development. Khalil et al. (2009) mentioned that a blood lead level higher than 8 μg/dl is responsible for increased mortality by coronary heart disease. Other authors Menke et al. (2006) found out that a blood lead level higher than 2 μg/dl was associated with myocardial infarction and stroke mortality. Although health related risks due to lead pipe poisoning were known from quite sometime and in fact many cities in the US tried to prohibit lead based plumbing in the 1920s, this was not enough to stop the counter actions of the lead pipe industry (Rabin 2008). In the 1970s the WHO still admitted 300 μg/l as the threshold for safe lead level in drinking water, but since then this value has fallen significantly (Table 2.4) as if all of a sudden the toxic risks of lead plumbing were made clear.

Some how this threshold evolution is quite similar to the asbestos problem in which an increased pattern of restrictions were adopted until, at last, the final prohibition came. It is then no surprise to see that the related Directive (98/83/CE) established a 15 year delay period before the 10 μg/l lead content threshold is enforced. This delay period is related to the cost of pipe substitution if the 10 μg/l lead content threshold was to be enforced immediately. A survey carried out in 1995 under the Directive (98/83/CE) revealed that Europe had almost 50 million meters of lead pipes. The replacement costs implied back then almost 34,000 million € (Papadopoulos 1999). More recent estimates points to 200,000 million € (Hayes 2009).

Table 2.4 Threshold evolution for lead content in drinking water in the last decades

Legal regulation	Year	Threshold for lead in drinking water (μg/l)
WHO	1970	300
Directive (80/778/CEE)	1980	50
Directive (98/83/CE)	From 25th December 2003 to 25th December 2013	25
	After 25th December 2013	10

2.9 Leaching and Eco-Toxicity Tests

As previously mentioned in Sect. 1.4, the eco-efficiency of construction and building materials encompasses waste recycling. Mortars, concretes and clay bricks constitute a viable way to reuse industrial wastes and further details on this subject will be carried out in Chaps. 5 and 6. Since many wastes contain heavy metals and other toxic substances it is then necessary to prove that they are safely encapsulated and do not harm the environment and the public health. Leaching is the process by which a contaminant is released from the solid phase to the waterphase. Leaching tests are used to assess that materials containing wastes in its composition do not constitute an environmental risk when in contact with surface or groundwater. It is impossible that a single leaching test can reproduce the conditions present in a real situation, therefore a wide variety of leaching tests exist (almost 50), but they represent minor variations of the same principles (Van Der Sloot et al. 1997). Table 2.5 presents some leaching tests.

At the European level there is no uniformity on leaching tests, however the European Committee for Standardization CEN TC 292—European Standardization of leaching tests, distinguishes three kinds of tests:

(a) Tests for basic characterization used to obtain the environmental performance of the waste in medium and long-term release of its constituents;
(b) Conformity tests used to verify if the waste meets certain benchmarks;
(c) On site verification tests, used to check in a fast manner if the waste are the same submitted to compliance tests.

The aforementioned tests assess the amount of contaminants present in the leachant solution being influenced by the leachant type (water or acid), temperature, specimen size, agitation degree, test period, liquid/solid ratio or the number of extractions. The choice of a particular leaching test should be done according to the type of waste and the immobilization process (Van Der Sloot 1996). For instance, Poon and Lio (1997) reported that the TLCP/EPA test is not suitable for measuring the immobilization of wastes in a cement matrix, because its alkalinity neutralizes the acidity of the leaching solution. Lewin (1996), argues that it is not realistic to expect that there is a test for each specific situation. Other authors

Table 2.5 Some leaching tests

Description	Year	Origin
EN 12457-3—characterization of waste—leaching-compliance leaching of granular waste material and sludges—Part 3: two stage batch test at a liquid to solid ratio of 2 and 8l/kg for materials with a particle size below 4 mm	2002	Europe
ASTM 6234/98—standard test method for shake extraction of mining waste by the synthetic precipitation leaching procedure	1998	US
NVN 7347—determination of the maximum leachable quantity and the emission of inorganic contaminants from granular construction materials and waste materials	1996	The Netherlands
AFNOR X-31-210/92 Essai de lexiviation.	1992	France
Toxic characteristics leaching procedure—TLCP/EPA	1992	US
NBR 10.005/1897—waste leaching	1987	Brazil
DIN 38414–S4—German Standard methods for the examination of water, waste water and sludge. Sludge and sediments (Group S). Determination of leachability (S4)	1984	Germany

(Hage and Mulder 2004) mentioned that the high variety of leaching tests is confusing and that the European committee for Standardization—CEN TC, got under way for several years, a project for its harmonization, which includes the development of new percolation test (pr EN14405), pH (CEN TC 292 WG6-N213) and a new batch test (prEN 12457). Although the toxicity of waste materials is usually assessed with leaching tests it should also be evaluated regarding their eco-toxicity using biologic tests. These tests assess the influence of the generated leachated in bacteria, algae growth, crustaceans mobility and even in the plant germination (Lapa et al. 2002). According to the standard procedures used in some European countries, if one of the two criteria (chemical or biological) is not met this means that the waste can not be directly used in the manufacture of building materials.

2.10 Conclusions

It has been shown throughout this chapter, that the toxicity of many building materials is a complex reality that implies a careful choice of them. Many materials used in the construction sector have some degree of toxicity, in terms of the environmental impacts of its production, by polluting indoor air or even by containing hazardous substances that imply a previous assessment. Unfortunately, the experience gained over the years shows that one cannot rely solely on legal regulations, as a guarantee that no risk to public health exist. Because the legal regulations are always a step behind of research findings and also because in some cases the legal regulations are more concerned to balance the public health and the economic impacts arising from setting very restrictive regulations. All this suggests that the toxicity of construction and building materials is an avoidable issue in the context of its eco-efficiency.

References

Akira M (2010) Asbestosis: IPF or NSIP-like lesions in asbestos-exposed persons, and such independency. Japanese J Chest Dis 69:38–44

Antonescu-Turcu A, Schapira R (2010) Parenchymal and airway diseases caused by asbestos. Curr Opin Pulmon Med 16:155–161. doi:10.1097/MCP.0b013e328335de61

ATSDR (2002) Toxicological profile for creosote. Agency for Toxic Substances and Disease Registry. U.S. Department of Health and Human Services, Public Health Sector, Atlanta

Axelson O, Fredrikson M, Akerblom G, Hardell L (2002) Leukemia in childhood and adolescence and exposure to ionizing radiation in homes built from uranium-containing alum shale concrete. Epidemiology 13:146–150

Azuma K, Uchiyama I, Chiba Y, Okumura J (2009) Mesothelioma risk and environmental exposure to asbestos: past and future trends in Japan. Int J Occup Environ Health 15:166–172

Baur X, Marek W, Ammon J (1994) Respiratory and other hazards of isocyanates. Int Arch Occup Environ Health 66:141–152. doi:10.1007/BF00380772

Berge B (2009) The ecology of building materials, 2nd edn. Architectural Press, Oxford

Bianchi C, Giarelli L, Grandi G, Brollo A, Ramani L, Zuch C (1997) Latency periods in asbestos-related mesothelioma of the pleura. Eur J Cancer Prev 6:162–166

Worldwatch Briefing (1998) Raw materials use and the environment. Worldwatch Institute, Washington

Bystrzejewska-Piotrowska G, Golimowski J, Urban P (2009) Nanoparticles: their potential toxicity, waste and environmental management. Waste Manag 29:2587–2595. doi: 10.1016/j.wasman.2009.04.001

Canfield R, Henderson C, Cory-Slechta D, Cox C, Jusko T, Lanphear B (2003) Intellectual impairment in children with blood lead concentrations below 10 μg per deciliter. N Engl J Med 348:1517–1526

Canut M (2006) Feasibility of using waste gypsum as a building material. Master's thesis, School of Engineering, Federal University of Minas Gerais

Chen J, Rahman N, Atiya I (2010) Radon exhalation from building materials for decorative use. J Environ Radioact 101:317–322. doi:10.1016/j.jenvrad.2010.01.005

Chester D, Hanna E, Pickelman B, Rosenman K (2005) Asthma death after spaying polyurethane truck bedliner. Am J Indust M 48:78–84. doi:10.1002/ajim.20183

Dhawan A, Sharma V, Parmar D (2009) Nanomaterials: a challenge for toxicologists. Nanotoxicology 3:1–9

Donaldson K, Poland C (2009) Nanotoxicology: new insights into nanotubes. Nat Nanotechnol 4:708–710. doi:10.1038/nnano.2009.327

Doroudiani S, Omidian H (2010) Environmental, health and safety concerns of decorative mouldings made of expanded polystyrene in buildings. Build Environ 45:647–654. doi: 10.1016/j.buildenv.2009.08.004

Dutrizac J, O'Reilly J, Macdonald R (1982) Roman lead plumbing: did it really contribuute to the decline and fall of the empire. CIM Bull 75:111–115

Edlich R, Winters K, Long W (2005) Treated wood preservatives linked to aquatic damage, human illness, and death-A societal problem. J Long-Term Effects Med Implant 15:209–223. doi:10.1615/JLongTermEffMedImplants.v15.i2.80

EPA (1992) Potential uses of phosphogypsum and associated risks. Office of Radiation Programs, 520/1-91-029, Washington

EURATOM (1996) Council directive 96/29 EC. European Atomic Comission, Brussels

Flores A, Ribeiro J, Neves A, Queiroz E (2004) Organochlorines: a public health problem. Environ Soc 7:111–124

Forsyth D, Jay B (1997) Organotin leachates in drinking water from chlorinated poly(vinyl chloride) (CPVC) pipe. Appl Organomet Chem 11:551–558. doi:10.1002/(SICI)1099-0739(199707)

Fristachi A, Xu Y, Rice G, Impellitteri C, Carlson-Lynch H, Little J (2009) Using probabilistic modeling to evaluate human exposure to organotin in drinking water transported by polyvinyl chloride pipe. Risk Anal 29:1615–1628. doi:10.1111/j.1539-6924.2009.01307.x

Gann R, Babrauskas V, Peacock R, Hall J (1994) Fire conditions for smoke toxicity measurements. Fire Mater 18:193–199. doi:10.1002/fam.810180306

Grassian V, O'Shaughnessy P, Adamcakova-Dodd A, Pettibone J, Thorne P (2007) Inhalation exposure study of titanium dioxide nanoparticles with a primary particle size of 2 to 5 nm. Environ Health Perspec 115:397–402. doi:10.1289/ehp.9469

Hage J, Mulder E (2004) Preliminary assessment of three new European leaching tests. Waste Manag 24:165–172

Hall J, Harwood B (1995) Smoke or burns—which is deadlier? N Fire Prot Assoc J 38:38–43

Hall J, Harwood B (1995) Smoke or burns—which is deadlier? N Fire Prot Assoc J 38:38–43

Hallock M, Greenley P, Diberardinis L, Kallin D (2008) Potential risks of nanomaterials and how to safe handle materials of uncertain toxicity. J Chem Health Saf 16:16–23. doi: 10.1016/j.jchas.2008.04.001

Hansen S, Burroughs H (1999) Classifying indoor air problems. Managing indoor air quality. Fairmont Press, Georgia, pp 62–63

Hauser R, Calafat A (2005) Phthalates and human health. Occup Environ Med 62:806–818. doi: 10.1136/oem.2004.017590

Hayes C (2009) Plumbosolvency control. Best practice guide. IWA specialist group on metals and related substances in drinking water. Cost 637. IWA Publishing, London

Heudorf U, Mersch-Sundermann V, Angerer J (2007) Phthalates: toxicolgy and exposure. Int J Hy Environ Health 210:623–634. doi:10.1016/j.ijheh.2007.07.011

Heuser V, Andrade V, Silva J, Erdtmann B (2005) Comparison of genetic damage in Brazilian footwear-workers exposed to solvent-based or water-based adhesive. Mutat Res 583:85–94. doi:10.1016/j.mrgentox.2005.03.002

Hirano S (2009) A current overview of health effect research on nanoparticles. Env Health Prev Med 14:223–225. doi:10.1007/s12199-008-0064-7

Hoch M (2001) Organotin compounds in the environment—An overview. Appl Geochem 16:719–743. doi:10.1016/S0883-2927(00)00067-6

Hodge A (1981) Vitrivius, lead pipes and lead poisoning. American J Archaeol 85:486–491

IARC (1995) IARC monographs on the evaluation of carcinogenic risks to humans. World Health Organization, International Agency For Research On Cancer, Lyon

IARC (1997) Polychlorinated dibenzo-para-dioxins and polychlorinated dibenzofurans. IARC monographs on the evaluation of carcinogenic risks to humans, vol 69. WHO, IARC, Lyon

ICRP (1990) Recommendations of the international commission on radiological protection. ICRP Publication 60, Pergamon Press, Oxford

Jarvholm B, Englund A, Albin M (1999) Pleural mesothelioma in Sweden: an analysis of the incidence according to the use of asbestos. Occup Environ Med 56:110–113. doi: 10.1136/oem.56.2.110

Khalil N, Wilson J, Talbottt E, Morrow L, Hochberg M, Hillier T, Muldoon S, Cummings S, Cauley J (2009) Association of blood lead concentrations with mortality in older women: a prospective cohort study. Environ Health 8:15. doi:10.1186/1476-069X-8-15

Koopman-Esseboom C, Weisglas-Kuperus N, De Ridder M, Van Der Paauw C, Tuinstra L, Sauer P (1996) Effects of polychlorinated biphenyl/dioxin exposure and feeding type on infants mental and psychomotor development. Pediatrics 97:700–706

Kostianien R (1995) Volatile organic compounds in the indoor air of normal and sick houses. Atmos Environ 29:693–702. doi:10.1016/1352-2310(94)00309-9

Kovler K (2009) Radiological constraints of using building materials and industrial by-products in construction. Constr Build Mater 23:264–253. doi:10.1016/j.conbuildmat.2007.12.010

Kovler K, Haquin G, Manasherov V, Ne'eman E, Lavi N (2002) Natural radionuclides in building materials available in Israel. Build Environ 37:531–537. doi:10.1016/S0360-1323(01)00048-8

Kwok N, Lee S, Guo H, Hung W (2003) Substrate effects on VOC emissions from an interior finishing varnish. Build Environ 38:1019–1026. doi:10.1016/S0360-1323(03)00066-0

Labat L, Olichon D, Poupon J, Bost M, Haufroid V, Moesch C, Nicolas A, Furet Y, Goullé J, Guillard C, Le Bouill A, Pineau A (2006) Variabilité de la mesure de la plombémie pour de faibles concentrations proches du seuil de 100 μg/l: étude multicentrique. Ann Toxicol 18:297–304
Ladou J (2004) The asbestos cancer epidemic. Environ Health Perspect 112:285–290. doi: 10.1289/ehp.6704
Lanting C, Patandin S, Fidlern V, Weisglas-Kuperus N, Sauer P, Boersma E, Touwen B (1998) Neurologic condition in 42-month-old children in relation to pre-and postnatal exposure to polychlorinated biphenyls and dioxins. Early Hum Dev 50:700–706. doi:10.1016/S0378-3782(97)00066-2
Lapa N, Barbosa R, Morais J, Mendes B, Méhu J, Oliveira J (2002) Ecotoxicological assessment of leachates from MSWI bottom ashes. Waste Manag 22:583–593. doi:10.1016/S0956-053X(02)00009-0
Levin B, Kuligowski E (2005) Toxicology of fire and smoke. In: Salem H, Katz S (eds) Inhalation toxicology, pp. 205–228 CRC Press, Boca Ratton
Lewin K (1996) Leaching tests for waste compliance and characterization: recent practical experiences. Sci Total Environ 178:85–94. doi:10.1016/0048-9697(95)04800-6
Liang H, Ho M (2007) Toxicity characteristics of commercially manufactured insulation materials for building applications in Taiwan. Constr Build Mater 21:1254–1261. doi: 10.1016/j.conbuildmat.2006.05.051
Littorin M, Truedsson L, Welinder H (1994) Acute respiratory disorder, rhinoconjuctivitis and fever associated with the pyrolysis of polyurethane derived from diphenylmethane diisocyanate. Scand J Work Environ Health 20:216–222
Liu A, Sun K, Yang J, Zhao D (2008) Toxicological effects of multi-wall carbon nanotubes in rats. J Nanoparticle Res 10:1303–1307. doi:10.1007/s11051-008-9369-0
Lovekamp-Swan T, Davis B (2003) Mechanisms of phthalate ester toxicity in the female reproductive system. Environ Health Perspect 111:139–145. doi:10.1289/ehp.5658
Marczynski B, Czuppom A, Hoffarth H, Marek W, Baur X (1992) DNA damage in human white blood cells after inhalation exposure to 4, 4'-methylenediphenyl diisocyanate (MDI)-case report. Toxicol Lett 60:131–138
Meeker J, Hu H, Cantonwine D, Lamadrid-Figueroa H, Calafat A, Ettinger A, Hernandez-Avila M, Loch-Caruso R, Téllez-Rojo M (2009) Urinary phthalate metabolites in relation to preterm birth in Mexico city. Environ Health Perspect 117:1587–1592. doi:10.1289/ehp.0800522
Menke A, Muntner P, Batuman V, Silbergeld E, Guallar E (2006) Blood lead below 0, 48 μmol/l (10 μg/dl) and mortality among US adults. Circulation 114:1388–1394. doi: 10.1161/CIRCULATIONAHA.106.628321
Morrell J (2002) Wood-based building components: what have we learned. Int Biodeterior 49:253–258. doi:10.1016/S0964-8305(02)00052-5
Nriagu J (1983) Saturnine gout among Roman aristocrats. Did lead poisoning contribute to the fall of the empire? N Engl J Med 308:660–663
Oppenhuizen A, Sijm D (1990) Bioaccumulation and biotransformation of poluchlorinated dibenzo-p-dioxins and dibenzofurans in fish. Environ Toxicol Chemistry 9:175–186
Paauw C, Tuinstra L, Sauer P (1996) Effects of polychlorinated biphenyl/dioxin exposure and feeding type on infants mental and psychomotor development. Pediactrics 97:700–706
Pacurari M, Castranova V, Vallyathan V (2010) Single and multi wall carbon nanotubes versus asbestos: are the carbon nanotubes a new health risk to humans. J Toxicol Environ Health 73:378–395. doi:10.1080/15287390903486527
Papadopoulos I (1999) Revision of the council directive on the quality of water intended for human consumption. Environmentalist 19:23–26. doi:10.1023/A:1006580705254
Papaefthmiou H, Gouseti O (2008) Natural radioactivity and associated radiation hazards in building materials used in Peloponnese, Greece. Radiat Meas 43:1453–1457. doi: 10.1016/j.radmeas.2008.03.032

Pavlidou S, Koroneos A, Papastefanou C, Christofides G, Stoulos S, Vavelides M (2006) Natural radioactivity of granites as building materials. J Environ Radioact 89:48–60. doi: 10.1016/j.jenvrad.2006.03.005

Peltonen K, Pfaffli P, Itkonen A, Kalliokoski P (1986) Determination of the presence of bisphenol-A and the absence of diglycidyl ether of bisphenol-A in the thermal degradation products of epoxy powder paint. Am Ind Hyg Assoc J 47:399–403. doi:10.1080/15298668691389946

Pocock S, Smith M, Baghurst P (1994) Environmental lead and children's intelligence: a systematic review of the epidemiological evidence. Br Med J 309:1189–1197

Poland C, Duffin R, Kinloch I, Maynard A, Wallace W, Seaton A, Stone V, Brown S, Macnee W, Donaldson K (2008) Carbon nanotubes introduced into the abdominal cavity of mice show asbestos-like pathogenecity in a pilot study. Nat Nanotechnol 3:423–428. doi:10.1038/nnano.2008.111

Poon C, Lio K (1997) The limitation of the toxicity characteristic leaching procedure for evaluating cement-based stabilised/solidified waste forms. Waste Manag 17:15–23. doi: 10.1016/S0956-053X(97)00030-5

Pruszinski A (1999) Review of the landfill disposal risks and the potential for recovery and recycling of preservative treated timber. Environ Prot Agency Rep, EPA

Rabin R (2008) The lead industry and lead water pipes "A modest campaign". Am J Public Health 98:1584–1592. doi:10.2105/AJPH.2007.113555

Rahman R (2007) PVC pipe and fittings: underground solutions for water and sewer systems in North America. 2nd Brazilian PVC Congress, Sao Paulo

Rihanek S (1971) Radioactivity of phosphate plaster and phosphate gypsum. Tonind-Ztg 95:264–270

Sadiki A, Williams D (1996) Speciation of organotin and organolead compounds in drinking water by gas chromatography-atomic emission spectrometry. Chemosphere 32:1983–1992. doi:10.1016/0045-6535(96)00097-5

Sadiki A, Williams D (1999) A study on organotin levels in Canadian drinking water distributed through PVC pipes. Chemosphere 38:1541–1548. doi:10.1016/S0045-6535(98)00374-9

Sadiki A, Williams D, Carrier R, Thomas B (1996) Pilot study on the contamination of drinking water by organotin compounds from PVC materials. Chemosphere 32:2389–2398. doi: 10.1016/0045-6535(96)00134-8

Salasar C (2007) Study on the emission of volatile organic compounds VOCs in house paints based on solvents based and in water. Master thesis, Londrina State University

Samfield M (1992) Indoor air quality data base for organic compounds. EPA-600/13

Satyanand T (2008) Aftermath of the Bhopal accident. Lancet 371:1900

Silverstein M, Welch L, Lemen R (2009) Developments in asbestos cancer risk assessment. Am J Indust Med 52:850–858

Singh N, Manshian B, Jenkins G, Griffiths S, Williams P, Maffeis T, Wright C, Doak S (2009) NanoGenotoxicology: the DNA damaging potential of engineered nanomaterials. Biomaterials 30:3891–3914. doi:10.1016/j.biomaterials.2009.04.009

Skarping G, Dalene M, Svensson B, Littorin M, Akesson B, Welinder H, Skerfving S (1996) Biomarkers of exposure, antibodies, and respiratory symptoms in workers heating polyurethane glue. Occup Environ Med 53:180–187

Smith P (2008) Risks to human health and estuarine ecology posed by pulling out creosote-treated timber on oyster farms. Aquatic Toxicol 86:287–298. doi:10.1016/j.aquatox.2007.11.009

Sterling D (1985) Indoor air and human healths. In: Gammage R, Kaye S, Jacobs V (eds) Volatile organic compounds in indoor air: an overview of sources, concentrations, and health effects. Lewis Publishers, USA

Stern B, Lagos G (2008) Are there health risks from the migration of chemical substances from plastic pipes into drinking water? A review. Hum Ecol Risk Assess 14:753–779. doi: 10.1080/10807030802235219

Swan S (2008) Environmental phthalate exposure in relation to reproductive outcomes and other health endpoints in humans. Environ Res 108:177–184. doi:10.1016/j.envres.2008.08.007

Tararbit K, Carré N, Garnier R (2009) Occurrence of lead poisoning during follow-up of children at risk wih initial screening lead blood levels below 100 µg/l. Revue d'Epidemiologie et de Sante Publique 57:249–255

Thierfelder T, Sandstrom E (2008) The creosote content of used railway crossties as compared with European stipulations for hazardous waste. Sci Total Environ 24:106–112. doi: 10.1016/j.scitotenv.2008.04.035

Thorton J (2000) Pandora's poison: chlorine, health, and a new environmental strategy. MIT Press, Cambridge

Thorton J (2002) Environmental impacts of polyvinyl chloride (PVC) building materials. University of Oregon, Cambridge

Tillitt D, Kubiak T, Ankley G, Giesy J (1993) Dioxin-like toxic potency in Forster's tern eggs form Green Bay, Lake Michigan, North America. Chemosphere 26:2079–2084. doi: 10.1016/0045-6535(93)90033-2

Troesken W (2006) The great lead water pipe disaster. MIT Press, Cambridge

Tsai W (2006) Human health risk on environmental exposure to bisphenol-A: a review. J Environ Sci Health-Part C Environ Carcinog Ecotoxicol Rev 24:225–255. doi:10.1080/10590500600936482

Tyshenko M, Krewski D (2008) A risk management framework for the regulation of nanomaterials. Int J Nanotechnol 5:143–160. doi:10.1504/IJNT.2008.016553

UNCHS (1997) Building materials and health. HS/459/97E. Habitat, Nairobi

Vale P, Rycroft J (1988) Occupational irritant contact dermatitis from fiberboard containing urea–formaldehyde resin. Contact Dermat 19:62. doi:10.1111/j.1600-0536.1988.tb02871.x

Van Der Sloot H (1996) Developments in evaluating environmental impact from utilization of bulk inert wastes using laboratory leaching tests and field verification. Waste Manag 16:65–81. doi:10.1016/S0956-053X(96)00028-1

Van Der Sloot H, Heasman L, Quevauviller P (1997) Harmonization of leaching/extraction tests. Elsevier, Amsterdam

Varma R, Mulay S (2006) The Bhopal accident and methyl isocyanate toxicity. Toxicol Organophosphate Carbonate Compd 7:79–88

Walker N, Bucher J (2009) A 21st century paradigm for evaluating the health hazards of nanoscale materials. Toxicol Sci 110:251–254

Wilbur S, Harris M, Cllure P, Spoo W (1999) Toxicology profile of formaldehyde. US Department of Health and Service DHHS, Public Health http://www.atsdr.cdc.gov/toxprofiles/tp.asp?id=220&tid=39. Accessed 10 September 2010

Wilhelm M, Dieter H (2003) Lead exposure via drinking water-unnecessary and avoidable. Umweltmedizin in Forschung und Praxis 8:239–241

Wolff M, Engel S, Berkowitz G, Ye X, Silva M, Zhu C (2008) Prenatal phenol and phthalate exposures and birth outcomes. Environ Health Perspect 116:1092–1097. doi:10.1289/ehp.11007

Wu C (2001) Discussion on fire safety factors from case studies of building fires. Master thesis, University of Taiwan

Wu W, Roberts R, Chung Y, Ernest W, Havlicek S (1989) The extraction of organotin compounds from polyvinyl chloride pipe. Arch Environ Contam Toxicol 18:839–843. doi: 10.1007/BF01160298

Yu Y, Zhang Q, Mu Q, Zhang B, Yan B (2008) Exploring the immunotoxicity of carbon nanotubes. Nanoscale Res Lett 3:271–277. doi:10.1007/s11671-008-9153-1

Zhang L, Steinmaus C, Eastmond D, Xin X, Smith M (2008) Formaldehyde exposure and leukemia: a new meta-analysis and potential mechanisms. Mutat Res 681:150–168. doi: 10.1016/j.mrrev.2008.07.002

Zietz B, Lab J, Dunkelberg H, Suchenwirth R (2009) Lead pollution of drinking water in lower saxony from corrosion of pipe materials. Gesundheitswesen 71:265–274

Chapter 3
Energy

3.1 General

The increasing demand for worldwide energy, is a major cause for the unsustainable development of our Planet. Between 2007 and 2030 energy demand will increase about 40% to 16.8 billion tons of equivalent petroleum-TEP (Fig. 3.1).

The rise in energy consumption has two main reasons, the increase in world population and the fact that there are an increasing number of people with access to electricity. Currently, 1.5 billion people still have no access to electricity (UN 2010). Beyond what energy consumption means in terms of using non-renewable fossil materials, the highest environmental impact of energy consumption, has to do with carbon dioxide emissions (Fig. 3.2), generated during the burning of coal and gas for electricity generation in power stations.

Given that buildings consume throughout their life cycle, more than 40% of all energy produced (OCDE 2003), we can easily see the high energy saving potential that this subsector may represent in terms of reducing carbon dioxide emissions.

3.2 Embodied Energy

The energy associated to construction and building materials (embodied energy), covers the energy consumed during its service life (Hammond and Jones 2008). There are however different approaches to this definition, namely: from the extraction of raw materials to the factory gate (cradle to gate), from extraction to site works (cradle to site) or from extraction to the demolition and disposal phase (cradle to grave). Berge (2009) considers as embodied energy, only the energy needed to put the material or product at the factory gate (first case), as to the transport energy and the energy related to the work execution both are included in the construction phase of the building. According to this author, the embodied energy represents 85–95% of the material total energy and the remaining 5–15%

F. P. Torgal and S. Jalali, *Eco-efficient Construction and Building Materials*,
DOI: 10.1007/978-0-85729-892-8_3, © Springer-Verlag London Limited 2011

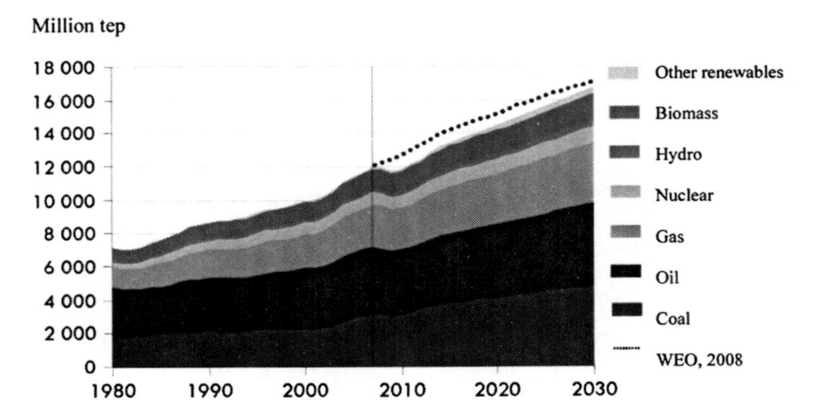

Fig. 3.1 World primary energy demand by fuel (WEO 2009)

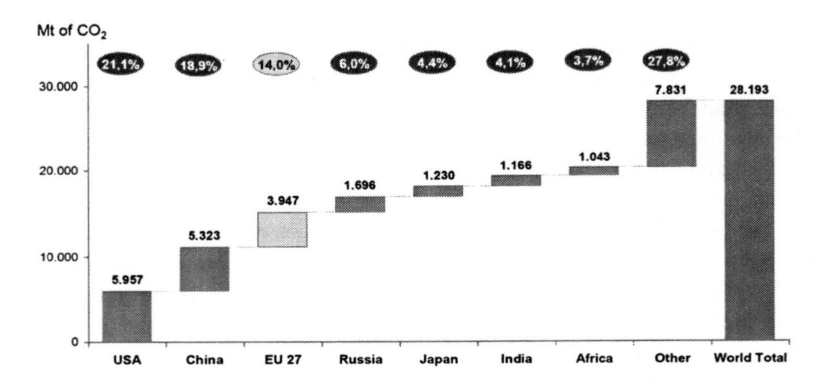

Fig. 3.2 World carbon dioxide emissions from the consumption of energy (ERA 2007)

Table 3.1 Energy consumption for construction materials (Reddy and Jagadish 2003)

Material transported	Energy consumption
Sand	1.75 (MJ/m^3 km)
Coarse aggregates	1.75 (MJ/m^3 km)
Bricks	2.0 (MJ/m^3 km)
Cement	1.0 (MJ/ton km)
Steel	1.0 (MJ/ton km)

relates to the construction, maintenance and demolition of the building. In the third case, the embodied energy includes all energy consumption phases from the production at cradle. As to the transport energy, this varies depending on the mode of transport: sea, air, road or rail. Reddy and Jagadish (2003) present some figures related to the transport energy of several construction materials (Table 3.1).

More recently, Berge (2009) presents some energy figures according to the transportation mode (Table 3.2).

Table 3.2 Transportation energy (Berge 2009)

Tranport mode	MJ/ton km
Plane	33–36
Highway (diesel)	0.8–2.2
Railway (diesel)	0.6–0.9
Railway (electricity)	0.2–0.4
Boat	0.3–0.9

Such a scenario shows the need to use local materials or to use materials away from the site, only if they have low density. Table 3.3 presents an inventory of the embodied energy of several construction materials used by several authors.

More recently Hammond and Jones (2008)presented the embodied energy and the embodied carbon for approximately 200 construction materials in the "cradle to gate" scenario.

This option intends to encourage a more rigorous establishment of the energy and carbon dioxide impacts of the transport phase. These authors based their study on the values used in the UK, but also on values currently used in Continental Europe. Morel et al. (2001) describe the construction of several houses in France, where it was possible to reduce the building energy by 215% with the use of local materials. Goverse et al. (2001) mentioned that an increased use of wood in buildings in Holland could reduce by almost 50% carbon emissions in that country.

Reddy and Jagadish (2003) observed that cement mortars have higher embodied energy than other mortars (Table 3.4). The same authors compared the embodied energy embedded in masonry, mentioning that the use of clay bricks, is not a wise option (Table 3.5).

According to Thormark (2006) an appropriate choice of construction and building materials can mean a 17% reduction in the energy of a building. Gonzalez and Navarro (2006) mentioned that a correct choice of building materials can reduce almost 30% CO_2 emissions, avoiding the emission of 38 tons of CO_2. These authors reported that a building constructed with materials of low environmental impact, will present CO_2 emissions of approximately 196 kg/m^2. For Dimoudi and Tompa (2008) the embodied energy of office buildings represent 13% to 19% of the operational energy for a service life of 50 years. Regarding the energy consumption and the carbon emissions per floor area, these authors report values of 1.93 GJ/m^2 and 198 kg CO_2/m^2 (building 1), for building 2 these values are 3.97 GJ/m^2 and 289.4 kg CO_2/m^2. The material differences between the two buildings are as follows: building 1 uses external walls made of double brick walls with a core thermal insulation, of a 5-cm thick extruded polystyrene layer and building 2 uses similar eexternal walls with a 5-cm thick mineral wool layer. The façade of building 1is covered by mortar while building 2 uses aluminum cladding. The flooring material in building 1 is ceramic tiles in the office spaces and marble in the corridors and staircases. Building 2 it uses vinyl tiles in the offices and marble in all corridors and staircases.These authors reported that the embodied energy in structural materials constitutes the most significant fraction, reaching 66.7% in the building 1 (42% for concrete and 24% for steel reinforcement) (Fig. 3.3).

Table 3.3 Embodied energy (Wellington 2005)

Material	MJ/kg	MJ/m^3
Agreggates	0.1	150
River aggregates	0.02	36
Extruded aluminum	201	542.700
Anodized extruded aluminum	227	612.900
Recycled aluminum	8.1	21.870
Recycled extruded aluminum	17.3	46.710
Recycled anodized aluminum	42.9	115.830
Asphalt	3.4	7.140
Bitumen	44.1	45.420
Cement	7.8	15.210
Cement mortar	2.0	3.200
Ready-mixed concrete (fc = 17.5 MPa)	1.0	2.350
Ready-mixed concrete (fc = 30 MPa)	1.3	3.180
Ready-mixed concrete (fc = 40 MPa)	1.6	3.890
Concrete block	0.94	–
Ceramic brick	2.5	5.170
Ceramic tile	0.81	–
Adobe blocks stabilized with cement	0.42	–
Compressed earth blocks (CEB)	0.42	–
Rammed earth stabilized with cement	0.8	–
Glass	15.9	40.060
Laminated glass	16.3	41.080
Stucco	4.5	6.460
Gypsum board	6.1	5890
Steel	32	251.200
Recycled steel	10.1	37.210
Local stone	0.79	1.890
Imported stone	6.8	1.890
Zinc	51	364.140
MDF	11.9	8330
Rough wood air-dried	0.3	165
Rough wood kiln-dried	1.6	880
Polished wood air-dried	1.16	638
Polished wood kiln-dried	2.5	1380
Plywood	10.4	–
Poliyester	53.7	7710
Poliuretane	74	44.400
PVC	70	93.620

Table 3.4 Embodied energy for several mortars (Reddy and Jagadish 2003)

Mortar	Cement	Soil	Sand	Energy per m^3 (MJ)
Cement	1	0	6	1,268
Cement + pozzolan	80%	0	6	918
Soil–cement	1	2	6	849
Lime–pozzolan (1:2)	–	0	3	732

Table 3.5 Embodied energy in masonry (Reddy and Jagadish 2003)

Masonry type	Energy per m^3 (MJ)	Energy percentage related to the masonry with ceramic bricks
Ceramic bricks	2,141	100
Cellular concrete blocks	1,396	65.2
Concrete blocks	819	38.3
Soil–cement blocks	646	30.2

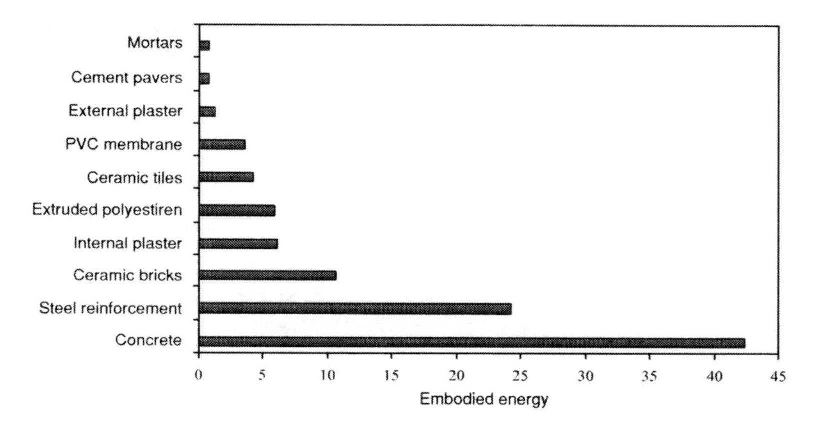

Fig. 3.3 Contribution of different materials into the embodied energy in building 1 (Dimoudi and Tompa 2008)

Berge (2009) compared several paving slabs (Fig. 3.4) and mentioned that although concrete tiles have a high embodied energy and thus high CO_2 emissions they can constitute a lesser environmental evil when compared to granite tiles with lower embodied energy and lower CO_2 emissions but with much higher transportation impacts.

The embodied energy of a C30 strength class concrete in Ireland is 1.08 MJ/kg (Goggins et al. 2010). These authors studied the energy saving potential associated with the replacement of Portland cement by blast furnace slag. They analyzed two types of concrete used for an office building in Ireland mentioning that when using concrete with 50% blast furnace slags they could save 924.175 MJ (Table 3.6).

In recent decades the operational energy in buildings (lighting, heating, cooling, etc.) was accepted to be the major part and that the embodied energy represented only a small fraction (10–15%). Consequently, much effort has been made towards the reduction of the operational energy by increasing the energy efficiency of the buildings. However, as operational energy is reduced the percentage of the embodied energy in the total energy of the buildings becomes increasingly prevalent. Thormark (2002) studied one of the buildings with the lowest energy consumption in Sweden (45 kWh/m^2) refering that the embodied energy for a lifetime of 50 years, could represent almost 45% of total energy. In a medium or even in a short term, it is possible that embodied energy could exceed the

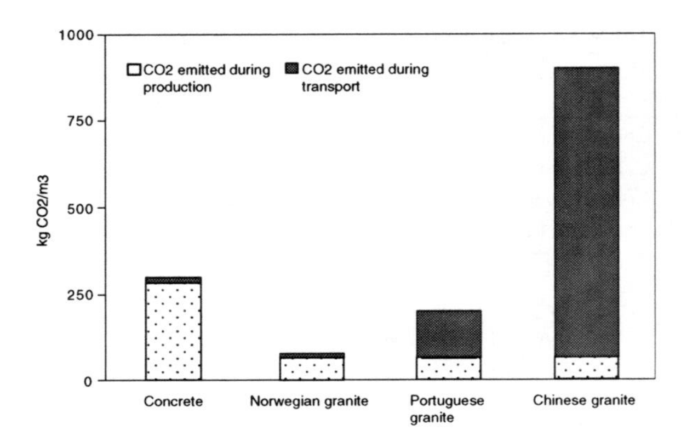

Fig. 3.4 CO_2 emissions associated with several flooring materials (Berge 2009)

Table 3.6 Embodied energy according to concrete composition (Gojjins et al. 2010)

Embodied energy (MJ)	Concrete without additions	Concrete with 50% BFS
Steel reinforcement	6.121	6.121
Aggregates	2.556	2.556
Cement	24.377	12.188
Water	15	15
Transport	428	428
Direct	7.358	7.358
Per slab panel	40.855	28.666
Total	3.064.125	2.139.950

operational energy in the building sector; in that scenario the embodied energy should be included in a future revision of the EU Directive on the Energy Performance of Buildings (Szalay 2007). More recently, Pacheco-Torgal and Jalali (2011) study a 97 apartment-type building (27.647 m^2) located in Portugal, concerning both embodied energy as well as operational energy. The operational energy was an average of 187.2 $MJ/m^2/year$ and the embodied energy accounted for aproximately 2,372 MJ/m^2, representing just 25.3% of the former. If the buildings were in the AA+ energy class it would mean that embodied energy could be almost four times the operational energy for a service life of 50 years.

3.3 Materials That Reduce Energy Consumption

3.3.1 Traditional Thermal Insulation Materials

Building insulation has been found to be one of the most cost-effective actions for the reduction of energy consumption and GHGs abatement (McKinsey and Company 2009). Thermal insulation materials represent a 21 billion € market share. These materials have a thermal conductivity factor, λ (W/m °C) lower than

Table 3.7 Thermal conductivity factor of some thermal insulation materials (Santos and Matias 2006)

Material	Dry density ρ (kg/m^3)	Thermal conductivity factor-λ-W/(m °C)
EPS-expanded polystyrene	>20	0.037
MW-mineral wool-rock	100–180	0.042
MW-mineral wool-glass	15–100	0.040
XPS-panels of extruded polystyrene	25–40	0.037
ICB-panels of expanded chipboard cork	90–140	0.045
PIR-rigid foam of polyurethane PUR-Rigid foam of poly-isocyanurate	20–50	0.040

Table 3.8 Thermal conductivity factor of some construction and building materials (Santos and Matias 2006)

Material	Dry density ρ (kg/m^3)	Thermal conductivity factor-λ-W/(m °C)
Aluminum	2,700	230
Steel	7,800	50
Granite	2,500–2,700	3.5
Soft limestone	1,600–1,790	1.1
Ceramic material	2,200–2,400	1.04
	<1,000	0.34
Ordinary concrete	2,300–2,600	2.0
Concrete with ligthweigth aggregates	400–600	0.24
Stucco	1,000–1,300	0.57
Ligthweigth wood	200–435	0.13
Adobe, CEB or rammed earth	1,770–2,000	1.1
Glass	2,200	1.4

0.065 and a thermal resistance higher than 0.30 (m^2. °C)/W (Santos and Matias 2006). Some materials like lightweight concrete or expanded clay do not respect these thresholds, however, if they have enough thickness they could be considered as thermal insulation materials. Regarding the masonry units (ceramic or concrete based) with improved thermal performance they will be discussed in Chap. 6. Thermal insulation materials can be as follows:

- Panels of expanded polystyrene (EPS)
- Panels of mineral wool (MW)
- Panels of extruded polystyrene (XPS)
- Panels of expanded chipboard cork (ICB)
- Rigid foam of poly-isocyanurate (PIR) or polyurethane (PUR)

Table 3.7 presents the thermal conductivity factor of the aforementioned thermal insulation materials. Table 3.8 presents the thermal conductivity factor of some construction and building materials for comparisons purposes.

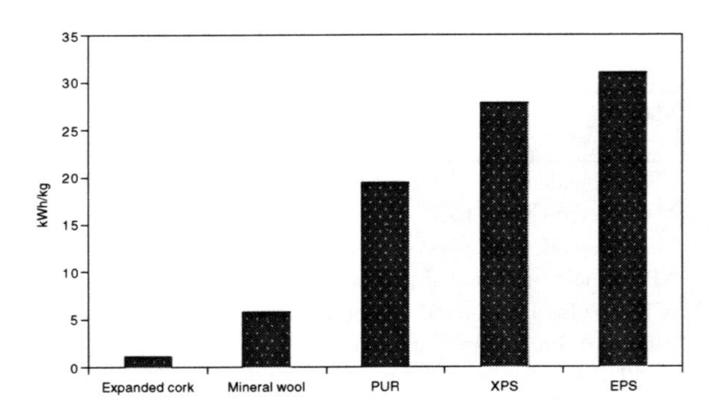

Fig. 3.5 Energy used in the production of several thermal insulation materials

Table 3.9 List prices of a packet of 100-mm thick thermal insulation sheet (Kymalainen and Sjoberg 2008)

Fibre raw material of insulation	Price (€/m^2)
Flax	10.8–15.4
Glass	5.4–8.7
Mineral	4.9–7.9
Wood fibre	9.3–11.8

The majority of the five thermal insulation materials previously described are associated with negative impacts in terms of their toxicity already discussed in Chap. 2. The only exception concerns the expanded cork, which is a product based on a renewable and completely recyclable material, cork. Even in terms of the energy consumed in the production phase, one can see that the expanded cork compares in a favourable manner with the other thermal insulation materials (Fig. 3.5). Besides, the cork production contributes to the preservation of cork trees, which are indispensable for the maintenance of the biodiversity in Alentejo, the Southern region of Portugal, the world's largest producer of cork.

3.3.2 Thermal Insulation Materials Based on Natural Materials

In recent years some investigations have focussed on thermal insulation materials based on natural materials like hemp fibres (Collet 2004). Although thermal insulation materials based on flax and hemp fibres show high insulation performance, these fibres are less cost-effective than glass or mineral fibres (Table 3.9). In addition, for a relative humidity above 80% they can be associated to molding.

More recently Collet et al. (2011) studied the water vapour properties of two hemp wools manufactured with different treatments. The first was constituted of hemp fibres linked with an organic binder and the second of hemp and cotton fibres with a polyester binder. According to these authors the results are useful to be used as input in numerical models in order to predict the hygrothermal behaviour of building walls.

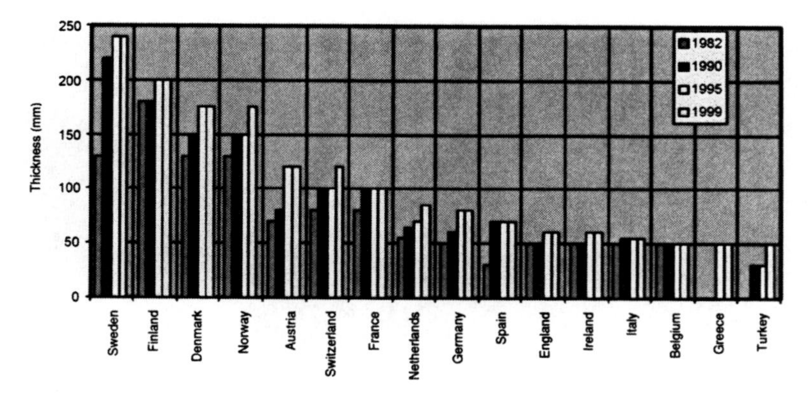

Fig. 3.6 Evolution of insulation thickness applicable in walls in Europe (Papadopoulos 2005)

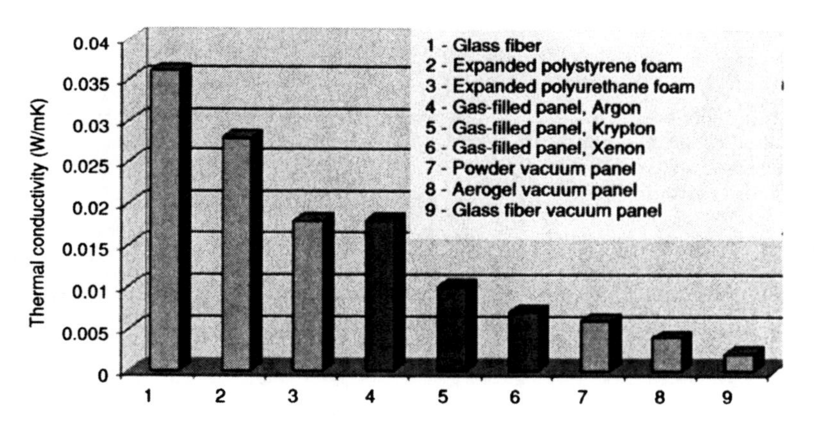

Fig. 3.7 Comparison of thermal conductivity between current insulation materials and high performance ones

3.3.3 High Performance Thermal Insulation Materials

The need to reduce energy costs in buildings, meant that the thickness of thermal insulation materials, has grown over the years, and in some countries of Northern Europe it has almost doubled (Fig. 3.6).

The development of thermal insulation materials with higher performance and lower thickness has become an important research field. At first, the solution involved the development of panels containing rare gases; however it soon became evident that their performance was surpassed by vacuum insulation panels (VIPs) which provide a thermal insulation almost ten times that of current thermal insulation materials (Fig. 3.7).

Fig. 3.8 Comparison between a VIP and a conventional thermal insulation material with the same thermal performance (Baetens et al. 2010)

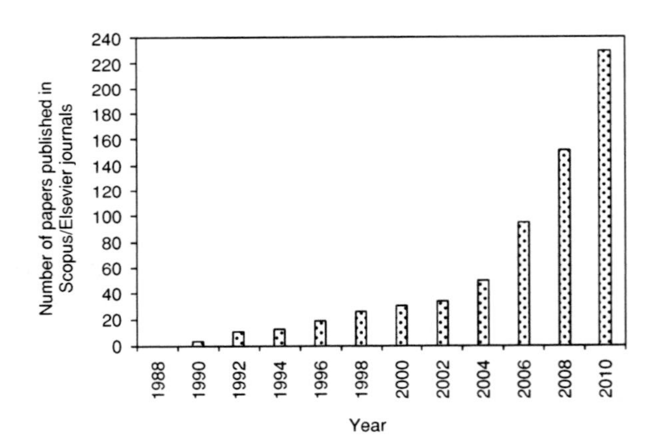

Fig. 3.9 Number of papers published in Scopus/Elsevier journals between 1988 and 2010 by the key words "PCM" or "phase change materials" in the title, abstract or keywords

VIPs consist of a core material, which is placed inside the vacuum panel and having a much lower thickness for the same thermal performance (Fig. 3.8).

Although the initial applications used a polystyrene core, lately they have been replaced by silica fume submitted to a compression up to 200 kg/m^3 which causes the air pore to be below the atmospheric pressure (Simmler and Brunner 2005). Some disadvantages of VIPs are the fact that they could not be cut on-site, its fragility brings the risk of being easily damaged and also the fact that they are associated to thermal bridging effects (Baetens et al. 2010). Fricke et al. (2008) mentioned that in Germany there are five companies dedicated to the production of vacuum insulation panels. Aerogel is another example of a high performance thermal insulation material. This material was developed by NASA in the 1950s

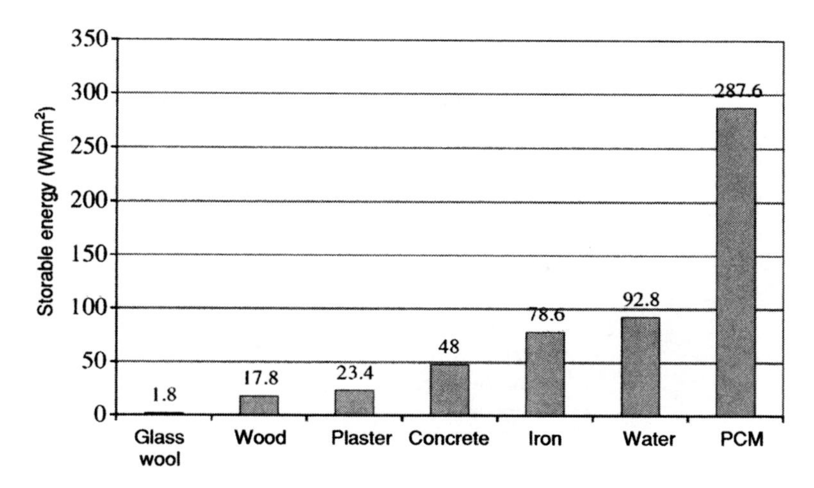

Fig. 3.10 Maximum storable energy between 18°C and 26°C for 10 mm of material and for 24 h (Kuznik et al. 2008, 2011)

and has been known as "solid smoke". It is composed of air (99.8%) and silica nanoparticles (0.2%) having the lowest thermal conductivity of any solid (between 0.004 and 0.03 W/mK). The current cost of this product is in the range of 25 €/m^2 which is almost 10 times higher when compared to a conventional insulation material for the same thermal resistance (Baetens et al. 2011). Some investigations have tried to produce aerogel-based windows (Wittwer 1992; Venkateswara et al. 2001; Baetens et al. 2011), this will allow for future high insulation windows representing its most valuable application in the near future.

3.3.4 Phase Change Materials

These materials use chemical bonds to store or release heat thus reducing energy consumption. Depending on the air temperature PCMs can change from solid to liquid or liquid to solid, absorbing or releasing heat during the process. Therefore, they can absorb heat inside buildings avoiding excessive heating or they can release heat in order to increase the minimum room temperature. Investigations related to PCMs started 30 years ago, but they have increase almost exponentially in the last five years (Fig. 3.9).

Early investigations used immersion processes and macro-capsules to integrate PCMs. These approaches have different drawbacks so PCMs had no big market impact. Recent advances in the technology of micro-encapsulation changed this situation (Schossig et al. 2005). PCMs can be organic, paraffin based or non-paraffin based (Table 3.10), they can also be inorganic like salt hydrate and

Table 3.10 Organic substances suitable for PCMs (Tyagi and Buddhi 2007)

Compound	Melting point (°C)	Heat of fusion (kJ/kg)
Butyl stearate	19	140
Paraffin C_{16}–C_{18}	20–22	152
Capric–lauric acid	21	143
Dimethyl sabacate	21	120
Polyglycol E600	22	127.2
Paraffin C_{13}–C_{24}	22–24	189
Mistiric acid (34%) + Capric acid (66%)	24	147.7
1-Dodecanol	26	200
Paraffin C_{18} (45–55%)	28	244
Vynil stearate	27–29	122
Capric acid	32	152.7

Table 3.11 Inorganic substances suitable for PCMs (Tyagi and Buddhi 2007)

Compound	Melting point (°C)	Heat of fusion (kJ/kg)
$KF \cdot 4H_2O$	18.5	231
$Mn(No3)_2\ 6H_2O$	25.8	125.9
$CaCl_2 \cdot 6H_2O$	29	190.8
$LiNO_3 \cdot 3H_2O$	30	296
$Na_2SO_4 \cdot 10H_2O$	32	251

metallics (Table 3.11), or even inorganic eutectics when PCMs are composed of two or more components which freeze and melt in a congruent manner (Table 3.12).

There are several ways of using PCMs in construction, including microcapsules, planar or cylindrical elements. PCM in the microcapsules is wrapped in a polymer coating and they are then mixed to mortars used in walls and ceilings. Table 3.13 shows several commercial PCMs recommended for building purposes (Tyagi and Buddhi 2007).

Athienitis et al. (1997) mentioned that the use of walls with PCMs allows a reduction of 4°C in the daytime maximum room temperature. Other authors (Darkwa and Kim 2004; 2005; Darkwa et al. 2006) observed that the random dispersion of the microcapsules of PCMs presents a performance reduction up to 20% than if PCMs were applied in laminated drywalls. Cabeza et al. (2007) used 5% of PCMs microcapsules in concrete panels 0.12 m thick, noticing an increase in the minimum room temperature of 2°C. As for the maximum room temperature, the same authors mentioned that when the outside temperature was 32°C, the concrete wall without PCMs reached 39°C, while the wall with this additive reached a maximum of 36°C. Alawadhi (2008) studied bricks with cylindrical holes filled with PCM reporting that the best results (a 18% reduction in the heat flux), occurred in bricks with three cylindrical holes placed at the centreline of the bricks. According to Kuznik et al. (2008, 2011) 1 cm is the optimal thickness for

Tabel 3.12 Eutectics substances suitable for PCMs (Tyagi and Buddhi 2007)

Compound	Melting point (°C)	Heat of fusion (kJ/kg)
66.6% $CaCl_2$ $6H_2O$ + 33.3% $Mgcl_2$ $6H_2O$	25	127
48% $CaCl_2$ + 4.3% NaCl + 0.4% KCl + 47.3% H_2O	26.8	188
47% $Ca(NO_3)_2$ $4H_2O$ + 53% $Mg(NO_3)_2$ $6H_2O$	30	136
60% $Na(CH_3COO)3H_2O$ + 40% $CO(NH_2)_2$	30	200.5

Table 3.13 Commercial PCMs (Tyagi and Buddhi 2007)

Designation	Substance	Melting point (°C)	Heat of fusion (kJ/kg)
RT20	Paraffin	22	172
Climsel23	Salt hydrate	23	148
Climsel24	Salt hydrate	24	216
RT26	Paraffin	25	131
RT25	Paraffin	26	232
STL27	Salt hydrate	27	213
S27	Salt hydrate	27	207
RT30	Paraffin	28	206
RT27	Paraffin	28	179
TH29	Salt hydrate	29	188
Climsel32	Salt hydrate	31	212
RT32	Paraffin	32	130

light panels containing PCMs. The results show that this thickness, allows for a doubled thermal inertia. Fig. 3.10 compares this solution with the energy storage capacity of several materials with 1 cm thickness.

These authors mentioned that 1 cm light panels with PCMs allows for a reduction of 4.2°C in the maximum room temperature. Kuznik and Virgone (2009) mentioned that the aforementioned light panels use 60 wt% of microencapsulated PCMs, while only 30 wt% PCMs can be incorporated in gypsum mortars. On the other hand when buildings reach to the end of their service life, gypsum mortars with PCMs cannot be considered an inert waste as happens with other mortars and this makes demolition and removal operations more complex.

3.4 Conclusions

Worldwide energy consumption is the main cause for the majority of carbon dioxide emissions. The European legislation related to the thermal efficiency of the building sector is responsible for a substantial reduction in energy consumption;

however, there is a limit beyond which no further reductions can be achieved within this legal framework. Thus, the use of building materials with lower embodied energy becomes a priority area. Most conventional thermal insulation materials, have disadvantages in terms of their toxicity, so the use of natural materials for insulation purposes can be a new step towards a more sustainable construction. Although the use of vacuum insulation panels, allows for high thermal insulation performance, it is not expected that in the short term their market share will rise in a sky rocketing manner because they are far from being cost-efficient when compared to conventional insulation materials. Translucent aerogels are much more prone to immediate commercial dissemination. Regarding PCMs, they have high potential in terms of mitigating the maximum and minimum room temperatures thus reducing energy demand without higher costs.

References

Alawadhi E (2008) Thermal analysis of a building brick containing phase change material. Energy Build 40:351–357. doi:10.1016/j.enbuild.2007.03.001

Athienitis A, Liu C, Hawe D, Hanu D, Feldman D (1997) Investigation of the thermal performance of a passive-solar test-room with wall latent-heat storage. Build Environ 32:405–410. doi:10.1016/S0360-1323(97)000097

Baetens R, Jelle B, Thue J, Tenpierik M, Grynning S, Uvslokk S, Gustavsen A (2010) Vacuum insulation panels for building applications: a review and beyond. Energy Build 42:147172. doi:10.1016/j.enbuild.2009.09.005

Baetens R, Jelle B, Gustavsen A (2011) Aerogel insulation for building applications: a state-of-the-art review. Energy Build 43:761–769. doi:10.1016/j.enbuild.2010.12.012

Berge B (2009) The ecology of building materials, 2nd edn edn. Architectural Press, Oxford

Cabeza L, Castellon C, Nogués M, Medrano M, Leppers R, Zubillaga O (2007) Use of microencapsulated PCM in concrete walls for energy savings. Energy Build 39:113–119. doi: #10.1016/j.enbuild.2006.03.030

Collet F (2004) Caracterisation hydrique et thermique des materiaux á faibles impacts environnementaux. Ph.D. Thesis, INSA, Rennes

Collet F, Achchaq F, Djellab K, Marmoret L, Beji H (2011) Water vapor properties of two hemp wools manufactured with different treatments. Constr Build Mater 25:1079–1085. doi: 10.1016/j.conbuildmat.2010.06.069

Darkwa K, Kim J (2004) Heat transfer in neuron-composite laminated phase-change drywall. J Power Energy Proc Inst Mech Eng 218:83–88. doi:10.1243/095765004773644085

Darkwa K, Kim J (2005) Dynamics of energy storage in phase-change drywall systems. Int J Energy Res 29:335–343. doi:10.1002/er.1062

Darkwa K, O'Callaghan P, Tetlow D (2006) Phase-change drywalls in a passive-solar building. Appl Energy 83:425–435. doi:10.1016/j.apenergy.2005.05.001

Dimoudi A, Tompa C (2008) Energy and environmental indicators related to construction of office buildings. Res Conserv Recycl 53:86–95. doi:10.1016/j.resconrec.2008.09.008

ERA (2007) A new energy ERA-Efficiency, renewables and clean thermal generation and advanced grid and strorage infrastructure. Vision paper for the EU strategic energy technology plan, Portuguese Ministery of Economy and Inovation. Lisbon

Fricke J, Heinemann U, Ebert H (2008) Vacuum insulation panels—From research to market. Vacuum 82:680–690. doi:10.1016/j.vacuum.2007.10.014

Goggins J, Keane T, Kelly A (2010) The assessment of embodied energy in typical reinforced concrete building structures in Ireland. Energy Build. 42:735–744. doi:10.1016/j.enbuild.2009.11.013

Gonzalez M, Navarro J (2006) Assessment of the decrease of CO_2 emissions in the construction field through the selection of materials. Build Environ 41:902–909. doi:10.1016/j.buildenv.2005.04.006

Goverse T, Kekkert M, Groenewegen P, Worrell E, Smits R (2001) Wood innovation in the residential construction sector: opportunities and constraints. Res Conserv Recycl 34:53–74. doi:10.1016/S0921-3449(01)00093-3

Hammond G, Jones C (2008) Inventory of carbon and energy (ICE) Version 1,6a. http:/www.bath.ac.uk/mech-eng/sert/embodied. Accessed November 2010

Kuznik F, Virgone J (2009) Experimental assessment of a phase change materials for wall building use. Optimization phase change material wallboard for building use. Appl Energy 86:2038–2046. doi:10.1016/j.apenergy.2009.01.004

Kuznik F, Virgone J, Noel J (2008) Optimization of a phase change material wallboard for building use. Appl Therm Eng 28:1291–1298. doi:10.1016/j.applthermaleng.2007.10.012

Kuznik F, Virgone J, Johannes K (2011) In situ study of thermal comfort enhancement in a renovated building equipped with phase change material wallboard. Renew Energy 36:1458–1462. doi:10.1016/j.renene.2010.11.008

Kymalainen H, Sjoberg A (2008) Flax and hemp fibers as raw materials for thermal insulations. Build Environ 43:1261–1269. doi:10.1016/j.buildenv.2007.03.006

McKinsey & Company (2009) Pathways to a low-carbon economy—Version 2 of the global greenhouse gas abatement cost curve. https://solutions.mckinsey.com/ClimateDesk/default.aspx. Accessed December 2010

Morel J, Mesbah A, Oggero M, Walker P (2001) Building houses with local materials: means to drastically reduce the environmental impact of construction. Build Environ 36:1119–1126. doi:10.1016/S0360-1323(00)00054-8

OCDE (2003) Environmental sustainable building—challenges and policies, Paris

Pacheco-Torgal F, Jalali S (2011) Embodied energy versus operational energy: a case study of a Portuguese 97 apartment-type building. Int J Sustain Eng (accepted)

Papadopoulos A (2005) State of the art in thermal insulation materials and aims for future developments. Energy Build 37:77–86. doi:10.1016/j.enbuild.2004.05.006

Reddy B, Jagadish K (2003) Embodied energy of common and alternative building materials and technologies. Energy Build 35:129–137. doi:10.1016/S0378-7788(01)00141-4

Santos C, Matias L (2006) Thermal coefficients for buildings envelope. Buildings ITE 50, LNEC, Lisbon

Schossig P, Hening H, Gschwander S, Haussmann T (2005) Micro-encapsulated phase-change materials integrated into construction materials. Solar Energy Mater Solar Cells 89:297–306. doi:10.1016/j.solmat.2005.01.017

Simmler H, Brunner S (2005) Vacuum insulation panels for building application basic properties, aging mechanisms and service life. Energy Build 37:1122–1131. doi:10.1016/j.enbuild.2005.06.015

Szalay A (2007) What is missing from the concept of the new European building directive. Build Environ 42:1761–1769. doi:10.1016/j.buildenv.2005.12.003

Thormark C (2002) A low energy building in a life cycle—its embodied energy, energy need for operation and recycling potential. Build Environ 37:429–435. doi:10.1016/S0360-1323(01)00033-6

Thormark C (2006) The effect of material choice on the total energy need and recycling potential of a building. Build Environ 41:1019–1026. doi:10.1016/j.buildenv.2005.04.026

Tyagi V, Buddhi D (2007) PCM thermal storage in buildings: a state of art. Renew Sustain Energy Rev 11:1146–1166

UN (2010) Energy for a sustainable future. The Secretary-General's Advisory Group on energy and climate change, New York

Venkateswara R, Pajonk G, Haranath D (2001) Synthesis of hydrophobic aerogels for transparent window insulation applications. Mater Sci Technol 17:343–348. doi:10.1179/026708301773002572

Wellington U (2005) Table of embodied energy coefficients. Centre for Building Performance. http://www.victoria.ac.nz/cbpr/projects/embodied-energy.aspx. Accessed October 2010

WEO (2009) World Energy Outlook.IEA. http://www.worldenergyoutlook.org/. Accessed December 2010

Wittwer V (1992) Development of aerogel windows. J Non-Cryst Solids 145:233–236. doi: 10.1016/S0022-3093(05)80462-4

Chapter 4
Construction and Demolition (C&D) Wastes

4.1 General

Although C&D wastes are a problem of increasing magnitude, there is little consensus about its volume. This subject is dependent on the absence of reliable statistics because in most countries these kinds of wastes are illegally dumped. Solis-Guzman et al. (2009) reported that worldwide, C&D wastes represent approximately 35% of the total waste and for Europe the same authors mention that C&D wastes represent 450 million ton/year. However, this figure cannot be taken for granted because it is unlikely that in Europe C&D wastes represent 22% of the total. For instance, the production of MSW ash exceeds C&D wastes more than 20 times (Tiruta-Barna et al. 2007). In the EU 15 the C&D wastes generated per capita are as much as 480 kg, meaning a total of 180 million ton/year. Kofoworola and Gheewala (2009) mentioned C&D wastes generated per capita for different European countries: Austria (300 kg), Denmark (500 kg), Germany (2,600 kg), The Netherlands (900 kg). The Eurostat (2010) mention a total of 970 million ton/year of C&D wastes and 2.0 ton/per capita. In terms of C&D wastes recycling rates, the values also differ from country to country. While the European average is only 25% (Solis-Guzman et al. 2009), some countries may reach 80%, as it happens in Denmark or in The Netherlands (Chini 2005). However, a recent report shows that these rates are very outdated (Table 4.1).

The landfill of C&D wastes generated in EU15 that are not recycled represent a volume with a 10 m height and 13 km^2 surface each year. The benefits of proper waste management are not solely environmental as we already saw in Chap. 1 but also economic. The preservation of biodiversity has a very relevant economic value associated with it. According to Weisleder and Nasseri 2006 the German market for recycled materials generated about 4,940 million euros in 2004, and the employment in this segment increased from 13,357 to 17,000 jobs between 2000 and 2004. A good example of the economic benefits associated with the recycling of C&D wastes is illustrated by the Environment Agency of the U.S. (EPA 2002), which states that the incineration of 10,000 tonnes of wastes can mean the creation

F. P. Torgal and S. Jalali, *Eco-efficient Construction and Building Materials*,
DOI: 10.1007/978-0-85729-892-8_4,

Table 4.1 Recycling rates of C&D wastes in Europe (Sonigo et al. 2010)

Countries	Recycling rates (%)
Belgium (Flanders)	Over 90
Denmark, Estonia, Germany, Ireland and The Netherlands	Over 70
Austria, Belgium, France, Lithuania, UK	60–70
Luxemburgo, Letónia, Eslovenia	40–60
Average recycling rate for EU-27	47
Cyprus, Czech Republic, Finland, Greece, Hungary, Poland, Portugal and Spain	Below 40
Bulgaria, Italy, Malta, Romania, Slovakia and Sweden	No data available

of one job, the landfill can create six jobs, but if the same amount of waste is recycled it can create 36 jobs.

4.2 Regulations

According to the Waste Management Acts 1996 and 2001, wastes can be defined as "any substance or object belonging to a category of waste which the holder discards or intends or is required to discard, and anything which is discarded or otherwise dealt with as if it were waste shall be presumed to be waste until the contrary is proved". The European waste catalogue-EWC encompasses 20 chapters related to different waste categories:

1. Wastes resulting from exploration, mining, quarrying, physical and chemical treatment of minerals;
2. Wastes from agriculture, horticulture, aquaculture, forestry, hunting and fishing, food preparation and processing;
3. Wastes from wood processing and the production of panels and furniture, pulp, paper and cardboard;
4. Wastes from the leather, fur and textile industries;
5. Wastes from petroleum refining, natural gas purification and pyrolytic treatment of coal;
6. Wastes from inorganic chemical processes;
7. Wastes from organic chemical processes;
8. Wastes from the manufacture, formulation, supply and use (MFSU) of coatings (paints, varnishes and vitreous enamels), sealants and printing inks;
9. Wastes from photographic industry;
10. Wastes from thermal processes;
11. Wastes from chemical surface treatment and coating of metals and other materials; non-ferrous hydro-metallurgy;
12. Wastes from shaping and physical and mechanical surface treatment of metals and plastics;
13. Oil wastes and wastes of liquid fuels (except edible oils, 05 and 12);

14. Waste organic solvents, refrigerants and propellants (except 07 and 08);
15. Waste packaging; absorbents, wiping cloths, filter materials and protective clothing not otherwise specified;
16. Wastes not otherwise specified in the list;
17. Construction and demolition wastes (including excavated soil from contaminated sites);
18. Wastes from human or animal health care and/or related research (except kitchen and restaurant wastes not arising from immediate health care);
19. Wastes from waste management facilities, off-site waste water treatment plants and the preparation of water intended for human consumption and water for industrial use;
20. Municipal wastes (household waste and similar commercial, industrial and institutional wastes) including separately collected fractions.

Table 4.2 presents Chap. 17 of the EWC related to the construction and demolition wastes.

Figure 4.1 presents a sequence to evaluate whether a waste can be classified as a hazardous one. According to Annex III of the European Council Directive 91/689/EC on hazardous waste, the properties of the wastes which render them hazardous are as follows:

H1: Explosive;
H2: Oxidizing;
H3: A—Highly flammable;
H3: B—Flammable;
H4: Irritant;
H5: Harmful;
H6: Toxic;
H7: Carcinogenic;
H8: Corrosive;
H9: Infectious;
H10: Teratogenic;
H11: Mutagenic;
H12: Substances and preparations which release toxic or very toxic gases in contact with water, air or an acid;
H13: Substances and preparations capable by any means, after disposal, of yielding another substance, e.g. a leachate, which possesses any of the characteristics listed above;
H14: Ecotoxic.

In 1991 the Japanese Government approved the "Recycling Law" under which they set minimum recycling targets for several by-products (Kawano 2003). As a consequence, the recycling percentages increase significantly (Fig. 4.2).

Since 1995 a C&D waste plan was implemented in Belgium. In 1996 the German industry accepted to cut by half the C&D wastes that were landfilled. In 1997 the Government of Finland decided that by the year 2000 50% of this waste

Table 4.2 Chapter 17 of the European waste catalogue-EWC

Code	Description
17	Construction and demolition wastes (including excavated soil from contaminated sites)
1701	Concrete, bricks, tiles and ceramics
170101	Concrete
17 01 02	Bricks
17 01 03	Tiles and ceramics
17 01 06	(*) Mixtures of, or separate fractions of concrete, bricks, tiles and ceramics containing dangerous substances
17 01 07	Mixture of concrete, bricks, tiles and ceramics other than those mentioned in 17 01 06
17 02	Wood, glass and plastic
17 02 01	Wood
17 02 02	Glass
17 02 03	Plastic
17 02 04	(*) Glass, plastic and wood containing or contaminated with dangerous substances
17 03	Bituminous mixtures, coal tar and tarred products
17 03 01	Bituminous mixtures containing coal tar
17 03 02	Bituminous mixtures containing other than those mentioned in 17 03 01
17 03 03	(*) Coal tar and tarred products
17 04	Metals (including their alloys)
17 04 01	Copper, bronze, brass
17 04 02	Aluminium
17 04 03	Lead
17 04 04	Zinc
17 04 05	Iron and steel
17 04 06	Tin
17 04 07	Mixed metals
17 04 09	(*) Metal waste contaminated with dangerous substances
17 04 10	(*) Cables containing oil, coal tar and other dangerous substances
17 04 11	Cables other than those mentioned in 17 04 10
17 05	Soil (including excavated soil from contaminated sites), stones and dredging spoil
17 05 03	(*) Soil and stones containing dangerous substances
17 05 04	Soil and stones other than those mentioned in 17 05 03
17 05 05	(*) Dredging spoil containing dangerous substances
17 05 06	Dredging spoil other than those mentioned 17 05 05
17 05 07	(*) Track ballast containing dangerous substances
17 05 08	Track ballast other than those mentioned in 17 05 07
17 06	Insulation materials and asbestos-containing construction materials
17 06 01	(*) Insulation materials containing asbestos
17 06 03	(*) Other insulation materials consisting of or containing dangerous substances
17 06 04	Insulation materials other than those mentioned in 17 06 01 and 17 06 03
17 06 05	(*) Construction materials containing asbestos

(continued)

Table 4.2 (continued)

Code	Description
17 08	Gypsum-based construction material
17 08 01	(*) Gypsum-based construction materials contaminated with dangerous substances
17 08 02	Gypsum-based construction materials other than those mentioned in 17 08 01
17 09	Other construction and demolition waste
17 09 01	(*) Construction and demolition wastes containing mercury
17 09 02	(*) Construction and demolition wastes containing PCB (for example PCB-containing sealants, PCB-containing resin-based floorings, PCB-containing sealed glazing units, PCB-containing capacitors)
17 09 03	(*) Other construction and demolition wastes (including mixed wastes) containing dangerous substances
17 09 04	Mixed construction and demolition wastes other than those mentioned in 17 09 01, 17 09 02 and 17 09 03

The wastes with (*) are considered hazardous wastes

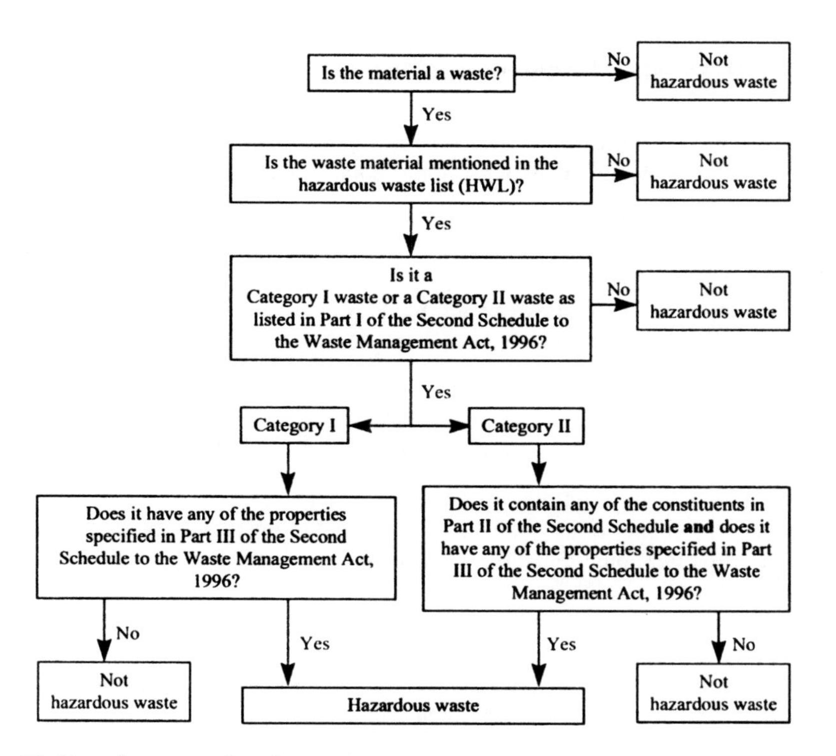

Fig. 4.1 Hazardous waste flowchart

should be recycled. In Spain, the first national plan on C&D waste occured in 2001. On November 19 of 2008 the EU approved the Revised Waste framework Directive No. 2008/98/EC. According to this Directive the minimum recycling percentage for

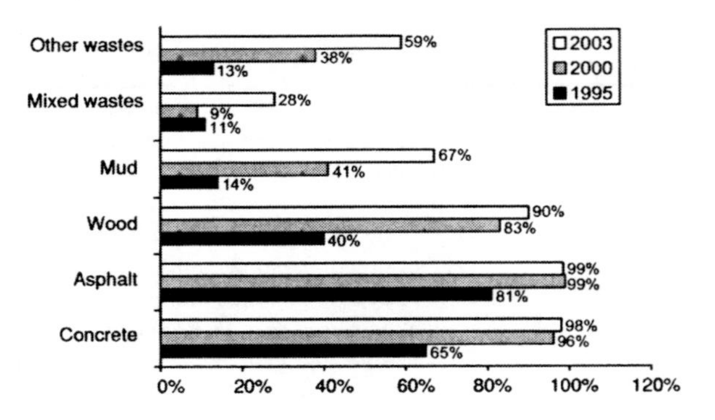

Fig. 4.2 Recycling percentage between 1995 and 2003 (Kawano 2003)

C&D wastes by the year 2020 should be at least 70% by weight. This regulation appears quite promising in order to achieve a more sustainable construction; however, it is unclear why it sets a delay of more than 10 years.

4.3 C&D Waste Management Plan

In order to reduce the C&D wastes several regulations impose the execution of a waste management plan. This plan should contain information about:

- Characterization of the construction works;
- Main waste streams;
- Waste management framework;
- Estimation of the quantities of each material;
- Proposal for minimization, reuse and recycling;
- Transport of the C&D wastes.

According to (DEHLG 2006) a C&D waste management plan should be prepared in case of development projects that exceed any of the following thresholds:

- New residential development of ten houses or more;
- New developments other than the above, including institutional, educational, health and other public facilities, with an aggregate floor area in excess of 1,250 m^2;
- Demolition/renovation/refurbishment projects generating in excess of 100 m^3 in volume, of C&D waste;
- Civil Engineering projects producing an excess of 500 m^3 of waste, excluding waste materials used for development works on the site.

Table 4.3 shows an example of a C&D waste management plan.

Tam (2008) ranked some measures to help the implementation of the C&D waste management plans:

1. Use of prefabricated building components
2. Purchase management

Table 4.3 C&D waste management plan (DEHLG 2006)

Project Name: [ ]

[Insert/Add/Delete to Detail as appropriate]

Description of Project:

The Project consists of the ______________(development/redevelopment etc.) of a ______________ (housing/ commercial/institutional/roads/water/wastewater etc.) scheme on a ______________ (greenfield/infill/redevelopment/ brownfield etc.) site. The project is situated at ________, ____________, Co. ________, in the administrative area of ______________ Council. The site of the works is located approximately ________ (metres/kilometres) from ________ (town/village/main road etc.) and access will be via the____________ (local/regional/national) road. The work will generally consist of the demolition of ___ (m3) of __________ and the construction of ________ (No./m2) of _________ (houses/offices/institutional/roads etc.).

In the course of the Project, it is estimated that the following quantities of C&D wastes/material surpluses will arise:

C&D Waste Material	Quantity (tonnes)
Clay and Stones	
Concrete	
Masonry	
Wood	
Packaging	
Hazardous Materials	
Other Waste Materials	
Total Arisings	

Table SF1: Estimated C&D Waste Arisings on Site

Proposals for Minimisation, Reuse and Recycling of C&D Waste

C&D waste will arise on the Project mainly from __________ (excavation/demolition) and ___________ (unavoidable construction waste/material surpluses/damaged materials). The ____________________ (Purchasing Manager etc.) shall ensure that materials are ordered so that the quantity delivered, the timing of the delivery and the storage is not conducive to the creation of unnecessary waste.

Excavated clay will be _________(carefully stored in segregated piles on the site for subsequent reuse/removed from site for direct beneficial use elsewhere). Concrete waste will be ____________ (source segregated/collected in receptacles with mixed C&D waste materials, for subsequent separation and recovery at a remote facility). Masonry and wood will be ____________ (source segregated/collected in receptacles with mixed C&D waste materials, for subsequent separation and recovery at a remote facility). Packaging will be ______________ (source segregated for recycling or return to suppliers). Hazardous wastes will be __________ (identified, removed and kept separate from other C&D waste materials in order to avoid further contamination). Other C&D waste materials will be ______________ (collected in receptacles with mixed C&D waste materials, for subsequent separation and disposal at a remote facility).

Excavation clay and C&D waste-derived aggregates are considered suitable for certain on-site construction applications. It is proposed that the following quantities, corresponding to all C&D Waste arisings from the project, will be used within the works:

(continued)

(continued)

Excavation clay and C&D waste-derived aggregates are considered suitable for certain on-site construction applications. It is proposed that the following quantities, corresponding to all C&D Waste arisings from the project, will be used within the works:

C&D Waste Type	Clay and Stones	Concrete	Masonry	TOTALS
Proposed Use	(t)	(t)	(t)	
Earthworks				
General Fill/Hardcore				
Pipe Bedding				
Selected Trench Backfill				
Fill to Structures				
Beneath Paths Structure				
Beneath Road Structure				
Other Site Use A				
Other Site Use B				
Off-Site Use				
TOTAL				

Table SF2: Proposals for Beneficial Use/Management of C&D Material Surpluses/Deficits and Waste Arisings on and off the Project

It is anticipated that waste materials ________(will/will not) have to be moved off site. It __(is/is not) the intention to engage specialist waste service contractors, who will possess the requisite authorisations, for the collection and movement of waste off-site, and to bring the material to a facility which currently (holds/does not hold) a ___________(Waste Licence/Waste Permit/Certificate of Registration). Accordingly, it will be necessary to arrange the following waste authorisations specifically for the Project:

Authorisation Type	Specific Need for Project (Yes/No?)	
Waste Licence	Yes	No
Waste Permit	Yes	No
Waste Collection Permit	Yes	No
Transfrontier Shipment Notification	Yes	No
Movement of Hazardous Waste Form	Yes	No

Table SF3: Specific Waste Authorisations Necessary for the Scheme

Demolition Procedures

The demolition works shall be undertaken in a manner which maximises the potential for recycling, including source segregating waste where appropriate. Activities shall be carried out in the following sequence:

Demolition Activity Sequence	General Description
Disconnection of Services/Vermin Control	Shutoff of E.S.B. , Gas etc.
Inventory of Hazardous Wastes	e.g. Asbestos etc.
Removal of Abandoned Furniture/Equipment	e.g. Furniture/White Goods
Removal of Asbestos/Hazardous Materials	e.g. Application of H&S Procedures
Removal of Fixtures	e.g. Fitted Presses etc.
Removal of Timber	e.g. Removal of Floors, Trusses, Rafters
Demolition of Structure Shell	Manual or Mechanical Demolition
Source Segregation of Material Fractions	Separation into Designated Material Fractions
Transport of Material from Site to Treatment Facilities	e.g. C&D Waste Recycling Facility
Transport of Material from Site to Controlled Disposal Sites	e.g. Inertised Hazardous Landfill Site
Site Preparation/Restoration	e.g. Hardstanding, Landscaping

(continued)

(continued)

Assignment of Responsibilities

A ______________ (Site Engineer/Manager/Assistant Manager etc.) shall be designated as the C&D Waste Manager and have overall responsibility for the implementation of the Project C&D Waste Management Plan. The C&D Waste Manager will be assigned the authority to instruct all site personnel to comply with the specific provisions of the Plan. At the operational level, a __________ (Ganger etc.) from the main contractor and ________ (appropriate personnel) from each sub-contractor on the site shall be assigned the direct responsibility to ensure that the discrete operations stated in the Project C&D Waste Management Plan are performed on an on-going basis.

Training

Copies of the Project C&D Waste Management Plan will be made available to all relevant personnel on site. All site personnel and sub-contractors will be instructed about the objectives of the Project C&D Waste Management Plan and informed of the responsibilities which fall upon them as a consequence of its provisions. Where source segregation, selective demolition and material reuse techniques apply, each member of staff will be given instructions on how to comply with the Project C&D Waste Management Plan. Posters will be designed to reinforce the key messages within the Project C&D Waste Management Plan and will be displayed prominently for the benefit of site staff.

Waste Auditing

The C&D Waste Manager shall arrange for full details of all arisings, movements and treatment of construction and demolition waste discards to be recorded during the construction stage of the Project. Each consignment of C&D waste taken from the site will be subject to documentation, which will conform with Table SF4 and ensure full traceability of the material to its final destination.

Detail	Particulars
Name of Project of Origin	e.g. New Harbour, Motorway
Material being Transported	e.g. Soil, Demolition Concrete, Crushed Asphalt etc.
Quantity of Material	e.g. 20.50 tonnes
Date of Material Movement	e.g. 01/01/2007
Name of Carrier	e.g. Authorised Carriers Ltd.
Destination of Material	e.g. Newtown Residential and Office Development
Proposed Use	e.g. Use as Hardcore in Dwelling Floors

Table SF4: Details to be Included within Transportation Dockets

Details of the inputs of materials to the construction site and the outputs of wastage arising from the Project will be investigated and recorded in a Waste Audit, which will identify the amount, nature and composition of the waste generated on the site. The Waste Audit will examine the manner in which the waste is produced and will provide a commentary highlighting how management policies and practices may inherently contribute to the production of construction and demolition waste. The measured waste quantities will be used to quantify the costs of management and disposal in a Waste Audit Report, which will also record lessons learned from these experiences which can be applied to future projects. The total cost of C&D waste management will be measured and will take account of the purchase cost of materials (including imported soil), handling costs, storage costs, transportation costs, revenue from sales, disposal costs etc. Costs will be calculated for the management of a range of C&D waste materials, using the format shown in Table SF5 below:

Material	Estimated Quantities & Costs (tonnes & Euro)
SOIL	
Quantity of Waste Soil (tonnes)	
Purchase Cost i.e. Import Costs (€)	
Materials Handling Costs (€)	
Material Storage Costs (€)	
Material Transportation Costs (€)	
Revenue from Material Sales (€)	
Material Disposal Costs (€)	
Material Treatment Costs (€)	
Total Waste Soil Management Costs (€)	
Unit Waste Soil Management Costs (€)	

Table SF5: Standard Record Form for Costs of C&D Waste Management (Sample relates to Soil – separate record forms should be compiled in respect of each waste material)

Table 4.4 Example of a calculation chart for Wambucalc (Lipsmeier and Gunther 2002)

Cat	Nr	Construction component	Construction design		SUM (kg)	Rate (%)
			Unit	Amount		
Shell						
S	1	Foundation	m^2	81.97	2,156.6	17.1

3. Education and training
4. Proper site layout planning
5. On-site waste recycling operation
6. Implementation of environmental management systems
7. High-level management commitment
8. Install underground mechanical wheel washing machines
9. Identification of available recycling facilitate
10. On-site sorting of construction and demolition materials

This author mentioned that the ''Low financial incentive'' and the ''Increase in overhead cost'' are considered the major difficulties in the implementation of the waste management plan. Other authors (Katz and Baum 2010) mentioned that on-site sorting of construction and demolition materials and waste management can slow down the construction rate.

Estimating C&D wastes. The estimation of the quantities of the different C&D wastes depends on the building system, the characteristics of the demolition and sorting process. It also depends on the amount of buildings under construction, rehabilitation or demolition at any given time, which will influence the quantities of C&D wastes for a certain area or country. Between 1998 and 2002 the University of Minho in Portugal participated in the European waste manual for building construction-WAMBUCO which was coordinated by the University of Dresden (Lipsmeier and Gunther 2002). In this project several card files were developed for specific waste building elements (walls, ceilings, floors and renders). Specific files were also developed for the amount of wastes associated with building construction. The specific card files allow for an estimation of the amount of wastes generated during the construction of a new building using the MS-Excel based Wambucalc software tool (Table 4.4).

The column "Sum" shows the total amount of waste generated during the specified construction component. The column "Rate" it shows the percentage contribution of the construction component on the overall waste generation. If the characteristics of building components are unknown it is still possible to estimate the quantity of waste produced on-site using the building card files. For this purpose it is necessary to know the construction area, the type of building (dwelling house, hotel or office building), and the comfort level (low, medium or high). Figure 4.3 shows the relationship between the floor area and the waste mass for dwelling houses.

Pascual and Cladera (2004) base their estimation of C&D wastes on the existence of a linear correlation between the amount of waste generation and the

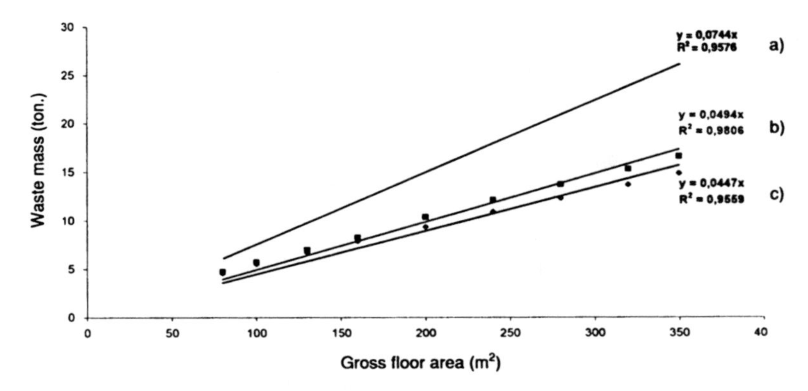

Fig. 4.3 Waste generation for dwelling houses according to the comfort level: **a** high; **b** medium; **c** low (Lipsmeier and Gunther 2002)

Table 4.5 Comparing the amount of C&D wastes generated during the construction phase (Mariano 2008)

Material	Amount of waste (kg/m^2)			
	Mariano (2008)	Monteiro et al. (2001)	Bohne et al. (2005)	Tozzi (2006)
Concrete	9.08	87	19.11	3.0
Ceramics	2.55		–	17.65
Mortar	2.93	189.0	–	18.33
Wood	16.82	3.0	2.75	0.87
Paper	0.16	21	0.46	0.58
Plastic	0.04			2.43
Fiber cement	0.63		0	
Others	1.94		6.19	
Hazardous	–	–	0.07	–
Gypsum	–	–	1.38	–
Glass	–	–	0.12	–
EPS	–	–	0.21	–
Metals	–	–	0.48	–
Total	34.15	300.0	30.77	42.89

consumption of Portland cement.Mariano (2008) analyzed the C&D wastes generated during the construction of a school with a floor area of 4,465 m^2, comparing it with that reported by other authors (Table 4.5)

This author mentioned that the large difference between the total value of 34.15 kg/m^2 and the value of 300 kg/m^2 reported by Monteiro et al. (2001), is due to the fact that in the first case some of the wastes have been reused and also because Monteiro also considered the demolition wastes. The differences between the results of Bohne et al. (2005) and Tozzi (2006) are minor and can be explained

Table 4.6 Efficiency of the waste management plan (Mariano 2008)

Material	Purchased materials (ton.)	Waste materials Q (ton.)	Efficiency (%)
Concrete	2,175.4	40.5	98.1
Ceramics	508.7	11.4	97.8
Mortar	629.8	13.0	97.9
Fibercement	27.6	2.8	89.8
Wood	88.3	75.1	14.9
Total	3,429.8	142.9	95.8

by the differences in the materials used in both cases. Mariano (2008) defines the efficiency of the waste management plan, as the relationship between the amount of materials purchased and the amount of waste generated at the end of the site work, which could not be reused (Table 4.6).

Although the value of the total efficiency is rather high, this form of accounting does not illustrate how much waste is recycled during the construction phase, which is a true measure of efficiency. In addition, this ratio does not allow for comparisons with smaller construction works in which the possibility of C&D wastes reuse is much lower. Solis-Guzman et al. (2009) mentioned that in February of 2008 a Government decree was enforced in Spain related to the C&D wastes management. This regulation requires the execution of a study about C&D wastes during the design phase. It also states that the contractor is responsible for the execution of a C&D wastes management plan. These two parts are required to obtain a building permit and must contain an estimation of the quantities of each waste stream and also an estimate of their treatment cost. The authors describe a new method for the estimation of C&D wastes whose indices were obtained from the study of a sample of 100 buildings. Lage et al. (2010) predict that in 2011 the Spanish region of Galicia will generate 2.2 million tonnes of C&D wastes per year, which corresponds to a ratio of 800 kg/person. In their study they assumed an estimate of 80 kg/m^2 of construction wastes from new constructions (0.11 m^3/m^2), an estimate of 80 kg/m^2 for renovation/rehabilitation works (without demolition) and an estimate of 1,350 kg/m^2 related to demolition wastes. Also, of the total C&D wastes produced, 40% relates to new construction, 20% to rehabilitation works and 40% to demolition works. Kofoworola and Gheewala (2009) used a value of 21.38 kg/m^2 as an estimate for the construction wastes generated in Thailand; this figure is quite low when compared with other countries. These authors mentioned that the reason is related to the fact that construction and maintenance of infrastructure (such as bridges and highways) were not considered. Demolition in any form was also not considered for the same reason. Last but not the least, most of the waste is dumped illegally and not accounted for. According to Katz and Baum (2010) the total amount of waste expected to accumulate exponentially in a residential construction site is estimated at 0.2 m^3/m^2 (Fig. 4.4).

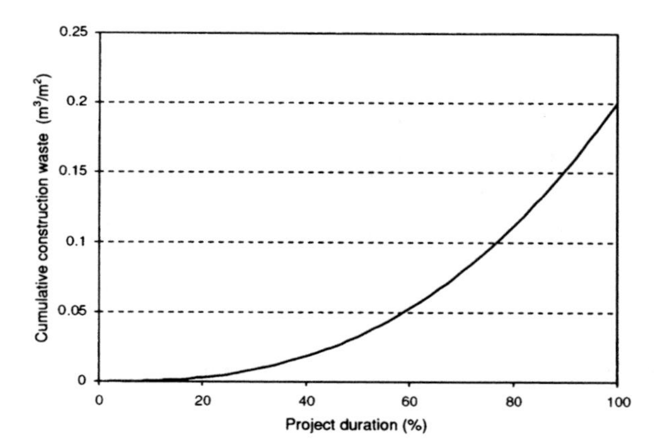

Fig. 4.4 Accumulation of construction waste throughout the project duration (Katz 2010)

4.4 Selective Demolition and Disassembly

Until very recently demolition processes were subordinate to a single principle, the minimization of the time spent in this operation, as a consequence the different waste streams would end all mixed up. However, the need to maximize the reuse and recycling of C&D wastes has forced the appearance of a new principle named "selective demolition" (Lipsmeier and Gunther 2002). The selective demolition involves the removal of components of the building in the inverse direction of its construction (Fig. 4.5).

Given that the selective demolition takes longer and is therefore more expensive than traditional demolition, this means that this technique could only be viable if financial compensation for this option is provided or if the regulations favour selective demolition. Regulations that set very low recycling rates, inhibit the implementation of selective demolition. Harnessing the full potential of selective demolition implies that in the design phase some principles to enhance the disassembly of the building are met (Kibert 2005):

1. Use of recycled and recyclable materials;
2. Minimize the number of types of materials;
3. Avoid toxic and hazardous materials;
4. Avoid composite materials and make inseparable products from the same material;
5. Avoid secondary finishes to materials;
6. Provide standard and permanent identification of material types;
7. Minimize the number of different types of components;
8. Use mechanical rather than chemical connections;
9. Use an open building system with interchangeable parts;
10. Use modular design;
11. Use assembly technologies compatible with the standard building practices;
12. Separate the structure from the cladding;
13. Provide access to all building components;

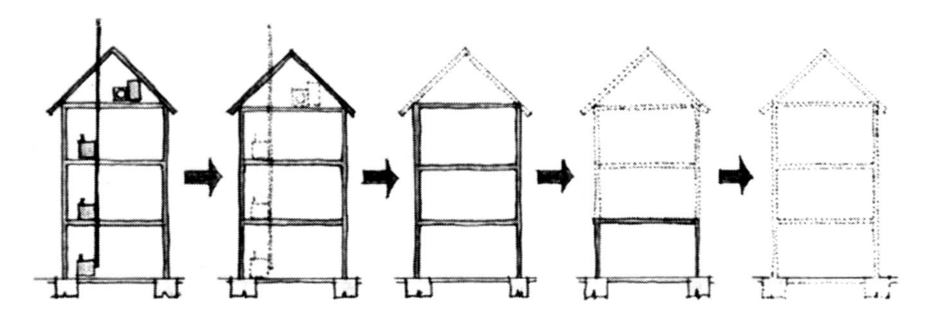

Fig. 4.5 Dismantling of a building in terms of selective demolition

14. Design components sized to suit handling at all stages;
15. Provide for handling components during assembly and disassembly;
16. Provide adequate tolerance to allow for disassembly;
17. Minimize the number of fasteners and connectors;
18. Minimize the types of connectors;
19. Design joints and connectors to withstand repeated assembly and disassembly;
20. Allow for parallel disassembly;
21. Provide permanent identification for each component;
22. Use a standard structural grid;
23. Use prefabricated assemblies;
24. Use lightweight materials and components;
25. Identify the point of disassembly permanently;
26. Provide spare parts and storage for them;
27. Retain information about the building and its assembly process.

According to Thormark (2007) design for disassembly has many environmental, economical as well as social benefits:

Economical motives

- Increased costs for waste handling
- Increased costs for extraction of resource
- Increased score in environmental labelling for demountable buildings
- Increased terminal value for demountable buildings

Social motives

- Demographic changes and changes in household structure
- Buildings are demolished before intended time

Environmental motives

- Increased problems with waste production
- Lack of virgin resources
- Recycling and the quality of the end products

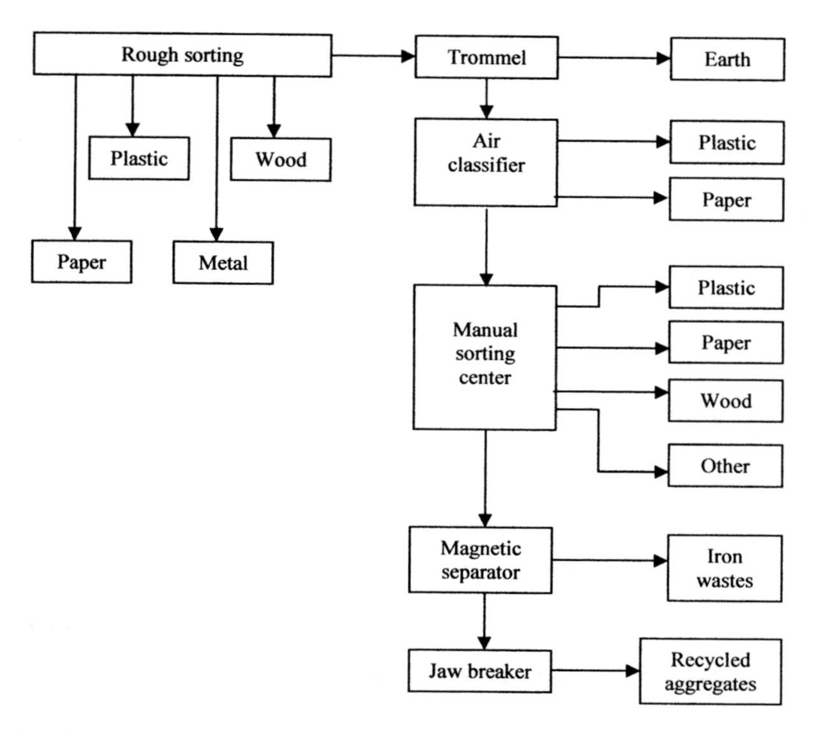

Fig. 4.6 Sorting process for production of recycled aggregates

- Reduced need of energy need for building operation
- Climate changes

4.5 On Site-Sorting and Recycling

On site-sorting allows for the separation of the different stream wastes (paper, wood, metal, plastic, etc.), being a crucial step to increase the recycling rate of C&D wastes. Wang et al. (2011) identified six critical success factors for on-site sorting of construction waste in China.

1. Manpower;
2. Market for recycled materials;
3. Waste sortability;
4. Better management;
5. Site space;
6. Equipment for sorting of construction waste.

The production of recycled aggregates from concrete and masonry wastes represents the most common case of recycling C&D wastes. The reuse of recycled

Table 4.7 Quality criteria for recycled gypsum (Demich 2008)

Parameter	Unit	Quality criteria (% mass)
Humidity	H_2O	<10
Calcium sulfate dihydrate	$C_aSO_4 \cdot 2H_2O$	>95
Chloride	Cl	<0.01
Soluble magnesium salts	MgO	<0.1
Soluble sodium salts	Na_2O	<0.06
Soluble potassium salts	K_2O	–
pH	–	5–9
Toxicity	–	Non toxic

aggregates in concrete is discussed in more detail in Chap. 5, nevertheless, it is important to draw attention to a crucial aspect that can influence the quality of recycled aggregates which has to do with the presence of undesirable materials such as soil, plastics, metals and organic matter. The sorting process can reduce this problem. Figure 4.6 shows the sorting process used for the production of recycled aggregates.

4.5.1 Recycling Gypsum-Based Materials

Gypsum materials can be recycled indefinitely without property loss. The recycling of gypsum materials requires grinding, removal of impurities and a low-temperature calcination. The construction activities directly related to the use of gypsum materials generate large amounts of wastes, either as renders for walls and ceilings or as plasterboards for drywalls. Although no specific values are known for the wastes generated by gypsum renders it is clear that a relevant part of these wastes are due to the fast hardening of the binder. According to Vanderley and Cincotto (2004) gypsum renders are associated with almost 45% of the wastes and plasterboards with about 10% to 12%. The use of recycled aggregates contaminated with gypsum particles for concrete production is a risk factor for concrete durability. The deterioration of concrete is caused by the chemical reaction of sulfate ions with the alumina of the aggregates or with the tricalcium aluminate (C_3A) of the hardened cement paste in the presence of water, both expansion products that can lead to the cracking of concrete. That is why the regulations on C&D wastes limited to less than 1% the presence of SO_3.

The recycling of gypsum plasterboards is already a consolidated reality in several countries. As long as the treatment process can reduce the amount of impurities present in the recovered gypsum this can be an important gypsum source. Table 4.7 defines the acceptance criteria for gypsum obtained from recycled plaster boards according to Eurogypsum.

Figure 4.7 shows the flowchart related to the recycling operations of gypsum plasterboards in Germany.

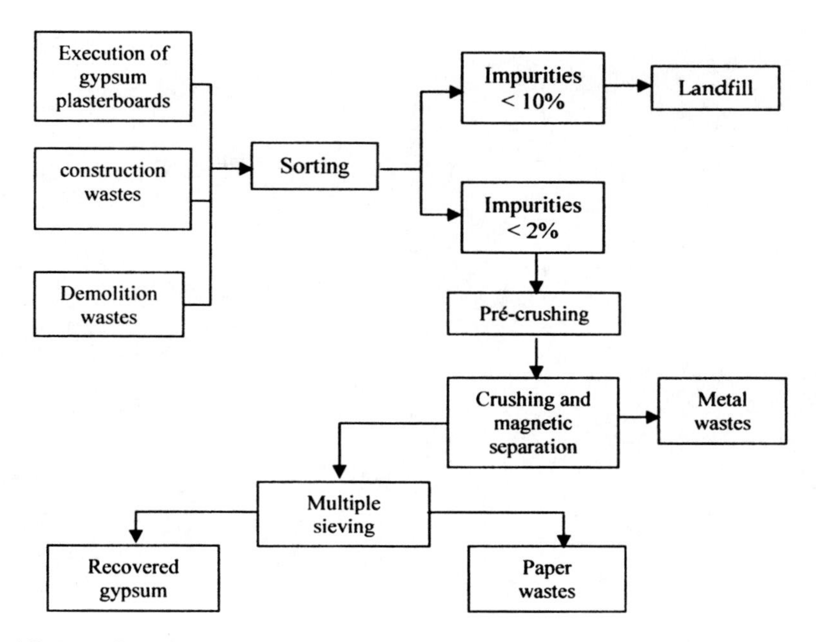

Fig. 4.7 Recycling of gypsum plasterboards. German process (Demich 2008)

4.5.2 Recycling Asbestos-Based Materials

Construction materials with asbestos are considered hazardous wastes under the European waste catalogue (code 170601 and code 170605), nevertheless, the state of the art about asbestos wastes points out to the possibility of their inertization and several industrial processes have already been developed for that purpose: INERTAM (Borderes 2000), ASBESTEX and ARI (Downey and Timmons 2005). The treatment of asbestos-based wastes can be as follows: Thermal treatments; chemical or mechanochemical treatments; microwave treatments (friable asbestos). Gualtieri and Tartaglia (2000) state that the use of a temperature treatment between 1,000 and 1,250°C allows the inertization of friable asbestos and also cement-based asbestos. The temperature is responsible for the transformations of the internal structure of asbestos into new and non-toxic crystalline phases (Fig. 4.8).

Different asbestos fibres present different performances when submitted to calcination operations (Table 4.8). Equations 1 and 2 show the transformations of chrysolite (1) and tremolite (2) asbestos as temperature increases:

$$\underset{\text{Chrysolite}}{Mg_3(OH)_4Si_2O_5} \xrightarrow{500\,^{\circ}C} \underset{\text{Metachrysolite}}{Mg_3Si_2O_7} + 2H_2O \xrightarrow{500\,^{\circ}C} \underset{\text{Forsterite}}{Mg_2SiO_4} + \underset{\text{Enstatite}}{MgSiO_3} \quad (1)$$

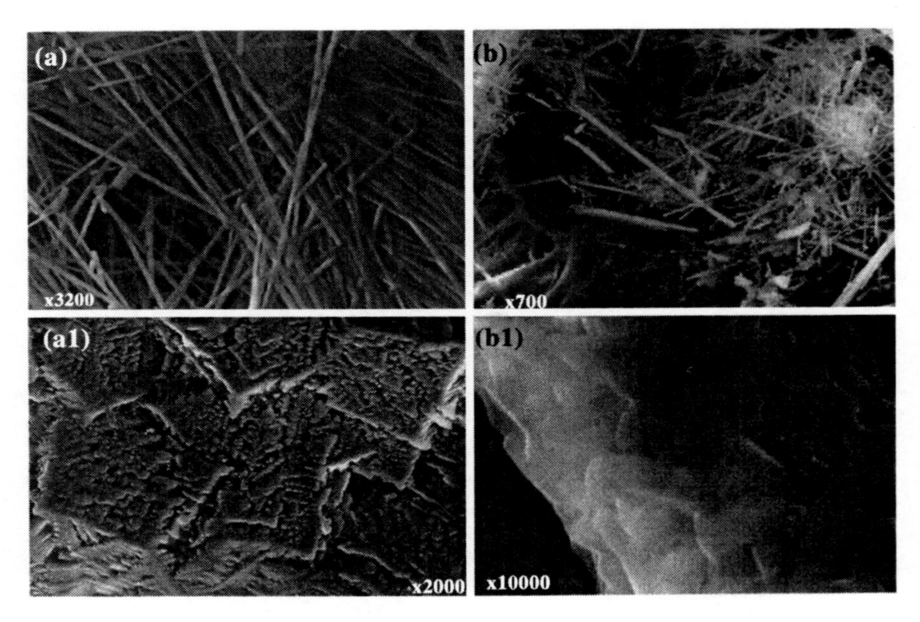

Fig. 4.8 Microstructure of asbestos fibres before and after thermal treatment: **a** and **a1** plain tremolite fibres; **b** and **b1** chrysolite fibres embedded in a cement matrix (Gualtieri and Tartaglia 2000)

$$\underset{\text{Tremolite}}{Ca_2Mg_5Si_8O_{22}(OH)_2} \xrightarrow{950\ ^{\circ}C} \underset{\text{Metatremolite}}{Ca_2Mg_5Si_8O_{23}} + 2H_2O \xrightarrow{1050\ ^{\circ}C} \underset{\text{Diopside}}{2CaMgSi_2O_6} + \underset{\text{Enstatite}}{3MgSiO_3} + \underset{\text{Chrystobalite}}{SiO_2} \qquad (2)$$

Leonelli et al. (2006) studied the inertization of friable asbestos and its incorporation as a magnesium source to produce ceramic-based products. They used a thermal treatment based on microwaves of 2.45 GHz during 13 min. The authors state that the cost of asbestos valorization varies between 0.05 a 0.2 €/kg, which is almost 10 times less than the cost of asbestos landfill disposal. Other investigations by Gualtieri et al. (2008a) confirm these results, referring to the possibility of using 3% to 5% of inertized asbestos in the production of porcelain products. Gualtieri et al. (2008b) patented a tunnel capable of achieving the inertization of asbestos cement wastes by using a temperature of 1,200°C during 16 h. This method has the advantage that it does not need the opening of the asbestos packing and does not require grinding operations. The authors use a low melting glass to reduce the calcination temperature. Dellisanti et al. (2009) refer to a pilot installation using the Joule vitrification method, in which a high intensity electric power (130 A) can melt the asbestos wastes at 1,500°C. Other authors (Plescia et al. 2003) used a mechanochemical treatment to change the morphology of the asbestos fibres into a

Table 4.8 Chemical and physical properties of common asbestos minerals (Leonelli et al. 2006)

Characteristic, chemical analysis (avg)	Serpentine, chrysotile $Mg_3Si_2O_5(OH)_4$	Amphibole				
		Crocidolite $Na_2Fe^{2+}3Fe_2^{3+}(Si_8O_{22})(OH)_2$	Amosite $(MgFe^{2+})7Si_8O_{22}(OH)_2$	Anthophyllite $Mg_7(Si_8O_{22})(OH)_2$	Tremolite $Ca_2Mg_5Si_8O_{22}(OH)_2$	Actinolite $Ca_2(MgFe^{2+})5Si_8O_22(OH)_2$
SiO_2	38–42	49–56	49–52	53–60	55–60	51–56
Al_2O_3	0–2	0–1	0–1	0–3	0–3	0–3
Fe_2O_3	0–5	13–18	0–5	0–5	0–5	0–5
FeO	0–3	3–21	35–40	3–20	0–5	5–15
MgO	38–42	0–13	5–7	17–31	20–25	12–20
CaO	0–2	0–2	0–2	0–3	10–15	10–13
Na_2O	0–1	4–8	0–1	0–1	0–2	0–2
Others	11.5–13	1.7–2.8	1.8–2.4	1.5–3	1.5–2.5	1.8–2.3
Decomposition temperature (°C)	450–700	400–600	600–800	600–850	950–1,040	620–960
Fusion temperature of residual material (°C)	1,500	1,200	1,400	1,450	1,315	1,400

non-toxic form. For friable asbestos Takahashi et al. (2009) reported a temperature treatment of 175°C during 24 h and the use of an NaOH (14 M) solution. Anastasiadoua et al. (2010) used a solution of acetic acid, a temperature range between 300 to 700°C and a pressure between 1.75 and 5.8 MPa. Zaremba et al. (2010) reported the detoxification of chrysotile asbestos through a low-temperature heating and grinding treatment with temperatures ranging from 500 to 725°C during 3h. Boccaccini et al. (2007) also used a microwave treatment to achieve the inertization of friable asbestos, turning fibrous structures into magnesium oxide blocks. The thermal treatment of asbestos cement wastes lead to a much lower toxicity level (Giantomassi et al. 2010). More recently Gualtieri et al. (2011) presented results on the reuse of calcined asbestos cement waste into brick, glass, plastics and pigments.

4.5.3 Recycling Concrete with a "Heating and Rubbing Method"

The recycling of concrete structures and masonry walls is carried out through various fragmentation operations (crushing and grinding). In order to reduce the dimensions of the concrete pieces jaw crushers, hammer mills and other mechanical devices are used. Although sorting operations can separate ceramic aggregates from concrete aggregates it is not easy to separate the rock fraction from the cement paste. Coarse aggregates with a cement paste have a higher water absorption reducing the performance of concrete (see Sect. 5.3). Moreover, assuming that the recycling plants will receive and process C&D wastes from numerous sources a dispersion of the properties of the recycled aggregates will increase leading to an increase in the dispersion of the quality of concrete. Shima et al. (2005) studied the possibility of submitting the concrete waste to a heat treatment in order to achieve a complete separation between the aggregates and the cement paste. They mentioned that using a heat treatment between 300 and 500°C allows obtaining aggregates identical to the original ones. The temperature rise causes the evaporation of the hydration water making the cement paste rather fragile, and the use of mechanical energy facilitates the separation between the aggregate and the binder. Mulder et al. (2007) used a similar procedure to separate the aggregates from the binder, however, they reported the need for a temperature near 700°C.

4.6 Conclusions

C&D wastes are a problem of increasing magnitude, however, reliable statistics are lacking because in most countries these kinds of wastes are illegally dumped. For instance there are no data on the recycling percentage of several European countries. Appropriate regulations on the recycling rates constitute a crucial step towards a more sustainable C&D wastes management. Several regulations impose the execution of a waste management plan, representing a crucial step towards C&D wastes reduction. The revised waste framework directive no. 2008/98/EC that sets the minimum recycling percentage for C&D wastes at least 70% by weight until the year 2020 will surely increase the recycling rate in the European area. The success of the C&D wastes recycling is also dependent on the demonstration of the economic advantages associated with it as happened with the techniques that made possible the inertization of asbestos-based materials. Quite appealing is the study of the EPA mentioning that C&D waste recycling allows the creation of a six-fold job increase than their disposal in a landfill. The use of selective demolition will increase the recycling rate of these wastes, and if disassembly principles were used during the design phase this will favour the selective demolition efficiency.

References

Anastasiadoua K, Axiotis D, Gidarakos E (2010) Hydrothermal conversion of chrysotile asbestos using near supercritical conditions. J Hazard Mater 179:926–932. doi:10.1016/j.jhazmat.2010.03.094

Boccaccini D, Leonelli C, Rivasi M, Romagnoli M, Veronesi P, Pellacani G, Boccaccini A (2007) Recycling of microwave inertised asbestos containing waste in refractory materials. J Eur Ceram Soc 27:1855–1858. doi:10.1016/j.jeurceramsoc.2006.05.003

Bohne R, Bergsdal H, Brattebo H (2005) Dynamic eco-efficiency modeling for recycling or C&D waste. Norwegian University of Science and Technology-Industrial Ecology Programme. http//:www.indecol.ntnu.no.AccessedJanuary2011

Borderes A (2000) Vitrification of the incineration residues. Revue Verre 6:1–2

Chini A (2005) Deconstruction and materials reuse—an international overview. CIB report TG 39, Publication 300, Rotterdam

DEHLG (2006) Best practice guidelines on the preparation of waste management plans for construction and demolition projects, Ireland

Dellisanti F, Rossi P, Valdré G (2009) Remediation of asbestos containing materials by Joule heating vitrification performed in a pre-pilot apparatus. Int J Miner Process 91:61–67. doi:10.1016/j.minpro.2008.12.001

Demich J (2008) Gypsum case study: recycling gypsum construction and demolition waste. The german model. In: Eurogypsum XVII congress, Brussels

Downey A, Timmons D (2005) Study into the applicability of thermochemical conversion technology to legacy asbestos wastes in the UK. In: WM′05 conference, Tucson, USA

EPA (2002) Resource conservation challenge: campaigning against waste. EPA 530-F-02–033

Giantomassi F, Gualtieri A, Santarelli L, Tomasetti M, Lusvardi G, Lucarini G, Governa M, Pugnaloni A (2010) Biological effects and comparative cytotoxicity of thermal transformed asbestos-containing materials in a human alveolar epithelial cell line. Toxicol in Vitro 24:1521–1531. doi:10.1016/j.tiv.2010.07.009

Gualtieri A, Tartaglia A (2000) Thermal decomposition of asbestos and recycling in traditional ceramics. J Eur Ceram Soc 20:1409–1418. doi:10.1016/S0955-2219(99)00290-3

Gualtieri A, Gualtieri M, Tonelli M (2008a) In situ ESEM study of the thermal decomposition of chrysotile asbestos in view of safe recycling of the transformation product. J Hazard Mater 156:260–266. doi:10.1016/j.jhazmat.2007.12.016

Gualtieri A, Cavenati C, Zanatto I, Meloni M, Elmi G, Gualtieri M (2008b) The transformation sequence of cement-asbestos slates up to 1,200°C and safe recycling of the reaction product in stoneware tile mixtures. J Hazard Mater 152:563–570. doi:10.1016/j.jhazmat.2007.07.037

Gualtieri A, Giacobbe C, Sardisco L, Saraceno M, Gualtieri M, Lusvardi G, Cavenati C, Zanatto I (2011) Recycling of the product of thermal inertization of cement–asbestos for various industrial applications. Waste Manag 31: 91–100. doi: 10.1016/j.wasman.2010.07.006

Katz A, Baum H (2010) A novel methodology to estimate the evolution of construction waste in construction sites. Waste Manag 31:353–358. doi:10.1016/j.wasman.2010.01.008

Kawano H (2003) The state of using by-products in concrete in Japan and outline of JIS/TR on recycled concrete using recycled aggregate. In Proceedings of the 1st FIB Congress on recycling, 245–253, Osaka

Kibert C (2005) Sustainable construction: green building design and delivery. Wiley, New York

Kofoworola O, Gheewala S (2009) Estimation of construction waste generation and management in Thailand. Waste Manag 29:731–738. doi:10.1016/j.wasman.2008.07.004

Lage I, Abella F, Herrero C, Ordonez J (2010) Estimation of the annual production and composition of C&D debris in Galicia (Spain). Waste Manag 30:636–645. doi: 10.1016/j.wasman.2009.11.016

Leonelli C, Veronesi P, Boccaccini D, Rivasi M, Barbieri L, Andreola F, Lancellotti I, Rabitti D, Pellacani G (2006) Microwave thermal inertisation of asbestos containing waste and its recycling in traditional ceramics. J Hazard Matrials B134:149–155. doi:10.1016/j.jhazmat.2005.11.035

Lipsmeier K, Gunther M (2002) WAMBUCO—waste manual for building constructions. Institute for Waste Management and Contaminated Sites Treatment of Dresden University of Technology, Dresden

Mariano L (2008) Management of construction wastes with structural reuse. Master thesis, University of Paraná, Curitiba

Monteiro J, Figueiredo C, Magalhães A, Melo M, Brito A, Brito J, Almeida T, Mansur G (2001) Manual for waste management. IBAM, Rio de Janeiro

Mulder E, De Jong T, Feenstra L (2007) Closed cycle construction: an integrated process for the separation and reuse of C&D waste. Waste Manag 27:1408–1415. doi:10.1016/j.wasman.2007.03.013

Pascual M, Cladera A (2004) Demolition waste management in Majorca: the particular case of an Island. In Proceedings of the internactional RILEM conference on the use of recycled materials in buildings structures, Barcelona

Plescia P, Gizzi D, Benedetti S, Camilucci L, Fanizza C, De Simone P, Paglietti F (2003) Mechanochemical treatment to recycling asbestos-containing waste. Waste Manag 23:209–218. doi:10.1016/S0956-053X(02)00156-3

Shima H, Tateyashiki H, Matsuhashi R, Yoshida Y (2005) An advanced concrete recycling technology and its applicability assessment through input–output analysis. J Adv Concr Technol 3: 53–67 www.j-act.org/headers/3_53.pdf

Solis-Guzman J, Marrero M, Montes-Delgado M, Ramirez-De-Arellano A (2009) A Spanish model for quantification and management of construction waste. Waste Manag 29:2542–2548. doi:10.1016/j.wasman.2009.05.009

Sonigo P, Hestin M, Mimid S (2010) Management of construction and demolition waste in Europe. Stakeholders Workshop, Brussels

Takahashi S, Ito H, Asai M (2009) Transformation of asbestos into harmless waste and recycle to zeolite by hydrothermal technique. J Soc Mater Sci Jpn 58:499–504

Tam V (2008) On the effectiveness in implementing a waste-management-plan method in construction. Waste Manag 28:1072–1080. doi:10.1016/j.wasman.2007.04.007

Thormark C (2007) Motives for design for dissassembly in building construction. In: Bragança L, Pinheiro M, Jalali S, Mateus R, Amoêda R, Correia Guedes M (eds) International congress sustainable construction, materials and practices challenge of the industry for the new millennium, Lisbon

Tiruta-Barna L, Benetto E, Perrodin Y (2007) Environmental impact and risk assessment of mineral wastes reuse strategies: review and critical analysis of approaches and applications. Resour Conserv Recycl 50:351–379. doi:10.1016/j.resconrec.2007.01.009

Tozzi R (2006) Characterization, evaluation and amangement of construction wastes. Master Thesis, University of Paraná, Curitiba

Vanderley J, Cincotto M (2004) Gypsum waste management alternatives. University of São Paulo, São Paulo

Wang J, Yuan H, Kang X, Lu W (2011) Critical success factors for on-site sorting of construction waste: a China study. Resour Conserv Recycl 54:931–936. doi:10.1016/j.resconrec.2010.01.012

Weisleder S, Nasseri D (2006) Construction and demolition waste management in Germany. ZEBAU, GmbH, Hamburg

Zaremba T, Krzakala A, Piotrowski J, Garczorz D (2010) Study on the thermal decomposition of chrysotile asbestos. J Therm Anal Calorim 101:479–485. doi:10.1007/s10973-010-0819-4

Chapter 5
Binders and Concretes

5.1 General

In the construction industry, the subject of binders and concretes represent the most substantial part of the consumption of non renewable raw materials, energy consumption and GHGs emissions. Concrete is the most used construction material on Earth, almost 10,000 million tons/year (Glavind 2009). The projections for the global demand of the main binder of concrete structures, Portland cement (Fig. 5.1) show that in the next 40 years the production of concrete will keep on rising.

Portland cement production represents 74% to 81% of the total CO_2 emissions of concrete, the aggregates represent 13% to 20%, therefore batching, transport and placement activities have no relevant expression in terms of carbon dioxide emissions (Flower and Sanjayan 2007). Figure 5.2 presents the flow diagram of Portland cement production. Portland cement CO_2 emissions result from the calcination of limestone ($CaCO_3$) and from combustion of the fossil fuels, including the fuels required to generate the electricity needed by the power plant. To make Portland cement clinker limestone is heated with a source of silica in a kiln at temperatures well over 1,350°C according to the reaction:

$$3CaCO_3 + SiO_2 \Rightarrow Ca_3SiO_5 + 3CO_2$$

The production of 1 ton of Portland cement generates 0.55 tons of chemical CO_2 and requires an additional 0.39 tons of CO_2 in fuel emissions for baking and grinding, accounting for a total of 0.94 tones of CO_2 (Gartner 2004).

The energy necessary for clinker production has declined gradually and substantially over the past 50 years, however, the current level of energy almost matches the theoretical minimum that ranges from 2,000 kJ/kg to 3,000 kJ/kg of clinker, so further reductions cannot be expected. Other authors (Damtoft et al. 2008) report that the cement industry emitted in 2000, on average, 0.87 kg of CO_2 for every kg of cement produced. Josa et al. (2007) used the 1992 CML methodology to assess the LCI of Portland cement produced in Holland, Switzerland,

F. P. Torgal and S. Jalali, *Eco-efficient Construction and Building Materials*,
DOI: 10.1007/978-0-85729-892-8_5,

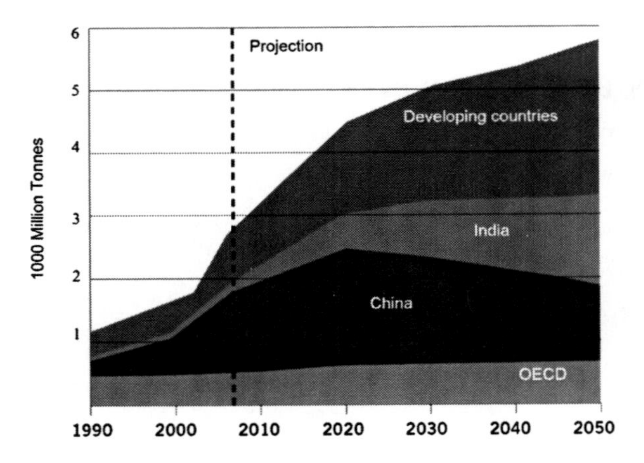

Fig. 5.1 Global cement demand by region and country (Taylor et al. 2006)

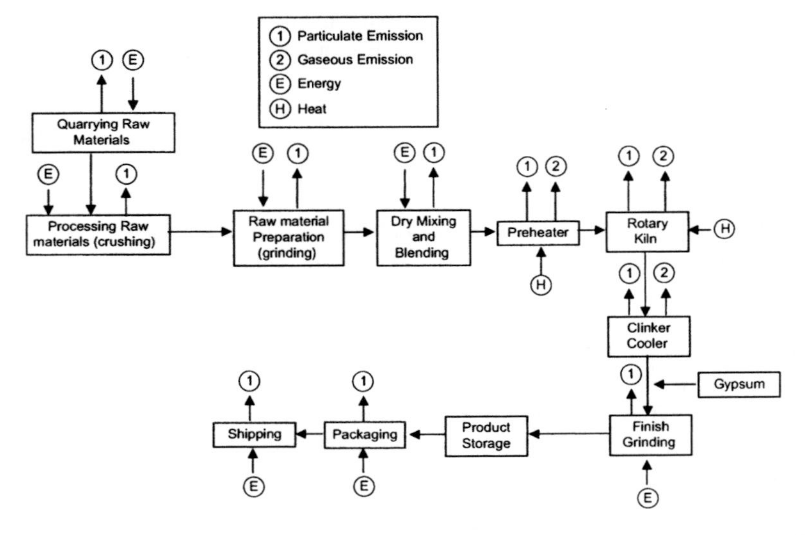

Fig. 5.2 Process flow diagram for the cement manufacturing process (Huntzinger and Eatmon 2009)

Sweden, Finland and Austria. The production of 1 kg of Portland cement can generate a maximum of 800 g of CO_2 in Type I cement. The same cement has SO_2 and NO_x emissions ranging from 1.1 g to 3.4 g of equivalent SO_2. However, a certain level of variability between different plants exists, Chen report almost 20% variations for the global warming impact (Chen et al. 2010). Regarding aggregates, their environmental impacts include non renewable raw materials consumption, energy consumption and more important the reduction of the biodiversity on the extraction sites. Besides, since the cost of aggregates are very dependent on the

Table 5.1 Production and reuse of some wastes in Europe (Titura-Barna et al. 2007)

Type of wastes	Production (Mt/ano)	Annual reuse (%)
MSWI	10.578	46
C&DW	420	0–90 (depends on the country)
Mineral wastes	400	–

transport distances it means that extraction operations very often locates near construction sites, thus multiplying the number of quarries and their biodiversity impacts. The worldwide consumption of aggregates is about 20,000 million tons/year and an annual growth rate of 4.7% is expected (Bleischwitz and Bahn-Walkowiak 2007). The partial replacement of Portland cement with pozzolanic wastes or the replacement of natural aggregates by industrial waste aggregates could play a key role in the eco-efficiency of construction and building materials (Meyer 2009). Due to their volume and characteristics some wastes deserve particular attention such as the ashes from the incineration of municipal solid waste (MSWI), C&DW and mineral wastes from mines and quarries (Table 5.1).

It is true that MSWI deserves some concern regarding the immobilization of their toxic substances and heavy metals which pose questions regarding the effectiveness of their immobilization (Tiruta-Barna et al. 2007), but that is not the case with C&DW and mineral wastes. C&DW represent 34% of the total wastes generated annually in the European area (excluding MSWI), being that some countries like Germany, Denmark and the Netherlands, have recycling rates reaching almost 90% (EDG 2004). As for the wastes from mines and quarries, they represent 27% of the total waste produced each year at the European level, though it is estimated that the mineral wastes already landfilled, amount to approximately 6,000 million tons (BRGM 2001). The reuse of wastes from mines and quarries in concrete not only reduces the consumption of natural aggregates but will also reduce the areas occupied by this type of waste (Yellishetty et al. 2008).

5.2 Concretes with Pozzolanic By-Products

The use of pozzolanic additions in construction dates back to several thousand years. Roy and Langton (1989) suggest that calcined clays mix with slaked lime (calcium hydroxide) were the first hydraulic binders made by men. Malinowsky (1991) reports about ancient constructions from 7000BC in the Galilei area (Israel) using this type of binder. The eruption of Thera in 1500 BC, which destroyed part of Santorini island was responsible for the appearance of large amounts of ashes used by the Greeks to make mortars that reveal having hydraulic properties. However, the Romans already knew that artificial pozzolans (calcined clay) were needed to produce mortars with a high performance, hence their use was not conditioned by the availability of natural pozzolans (Hazra and Krishnaswamy 1987). The Roman mortars used for the Hadrian's wall in Britain were made of crushed ceramic material mixed with lime binder (Guleç and Tulun 1997). Crushed ceramics seem also to be preferred from early Hellenistic to early Byzantine times in mortars related to water-bearing constructions and to protect the insides of walls from moisture, typically in baths, canals and aqueducts (Sbordoni-Mora 1981; Degryse et al. 2002). Several of the monuments that survived till the twenty-first century like the triumphal arches of the Emperors Claudius and Trajan in Ostia or the bridges of Fabricus, Aemilius, Elius e Milvius show the durability of lime-pozzolan based mortars (Bogue 1955; Lea 1970). Some results of mortar samples from 7000 BC reveal a dense material having a compressive strength of almost 30 MPa (Bentur 2002). However, the appearance of Portland cement in the nineteenth century, having a fast setting and higher early strength was responsible for the decline of the use of lime-pozzolan binders. Despite recent advances in kiln design and alternative, low energy clinkers, it seems likely that the greatest carbon savings from the industry are likely to be made by the inclusion of supplementary cementing materials (Mehta 2001; Roskovic and Bjegovic 2005; Bargaheiser and Nordmeyer 2008; Palmer 2010; Tyrer et al. 2010). Besides, reusing pozzolanic by-products also avoids the use of large areas for landfill disposal which is a major threat for biodiversity.

5.2.1 Pozzolans, Pozzolanic Reaction and Pozzolanic Activity

Several standards define pozzolans as siliceous and aluminous materials which have very little or no cementitious characteristics but when finely divided in the presence of water react with calcium hydroxide to form cementitious compounds (ASTM 2007). The pozzolanic activity is a rather complex property which relies on the amorphous state of silica and aluminum, being higher with higher amorphous states (Gjorv 1992). Generally speaking the aluminosilicate species of the pozzolans will react with calcium hydroxide to form calcium-silico aluminate phases (Papadakis et al. 1992):

- $2CaO{\cdot}Al_2O_3.SiO_2{\cdot}8H_2O$;
- CSH;
- Calcium aluminate hydrates.

Pozzolans can be of natural origin or artificial like calcined clays or industrial by-products (Neville 1997; Mehta 1998). Natural pozzolans came from silicon rich magma that solidified very rapidly remaining in an amorphous state. As for artificial pozzolans, clays submitted to a calcination process below the dehydroxylation temperature (Al-Rawas et al. 1998) they became structurally instable because of the hydroxyl groups that left out due to the calcination process (Menezes et al. 2003). The pozzolanic activity of calcined clays is very much dependent on the loss of structural water which favours the creation of an amorphous structure (Ambroise et al. 1986). Some authors (Ambroise et al. 1985) compared several clays calcined at 750°C observing that kaolinite clays were the

Table 5.2 Effect of calcination on the specific surface of kaolinite, illite and montmorillonite (Fernandez et al. 2011)

Clay	Specific surface (m^2/g)
Raw kaolinite	26.1512
Kaolinite 600°C	24.698
Kaolinite 800°C	24.1283
Raw illite	21.3277
Illite 600°C	18.4316
Illite 800°C	13.3214
Raw montmorillonite	31.0287
Montmorillonite 600°C	21.386
Montmorillonite 800°C	9.7221

most reactive, and that illite and montmorillonite have low pozzolanic activity. As for muscovite and phlogopite, they show no pozzolanic activity. He et al. (1995a, b, 2000) also studied the influence of thermal treatment on the pozzolanic activity of several clays reporting different behaviors for different clays. The high pozzolanic activity of metakaolin is confirmed by other authors (Oliveira et al. 2005; Barbhuiya et al. 2009). Fernandez et al. (2011) compared (kaolinite, illite and montmorillonite) when thermally treated at 600°C and 800°C referring that kaolinite have a high pozzolanic activity because the location of the hydroxyl groups in the kaolinite structure favours more disorder and exposure of Al groups at the surface of the material. For the pozzolanic industrial by-products such as fly ash or silica fume a similar process occurs because these materials have a very high content of silicon and aluminum (Agarwal 2006). The main factors constituting a good pozzolanic activity also include small particle size and high specific surface area. Since the pozzolanic reaction takes place between the surface of pozzolan particles and the calcium hydroxide the surface characteristics will influence the pozzolanic activity. Therefore, pozzolans with high specific surfaces are needed to maximize the surface area that will react with the calcium hydroxide. Some authors (Coutinho 1988) mention the need for pozzolans with a minimum Blaine fineness of 3,000 cm^2/g, although Blaine fineness values between 6,000 and 10,000 cm^2/g should be used. Day and Caijun (1994) mention that increasing the Blaine fineness from 2,500 to 5,500 cm^2/g increases the pozzolanic activity mostly at the early ages. Increasing the pozzolan fineness increases its strength activity, however, it also increases the water requirements so that there is an optimum particle size that maximizes the strength behaviour (Bouzoubaa et al. 1997). Some authors (Kiattikomol et al. 2001) mentioned that the fineness of fly ash is more important than its chemical composition when determining the strength activity. For some clays, the thermal treatment leads to an agglomeration behaviour that reduces the specific surface found to be the highest for montmorillonite, followed by illite and kaolinite (Table 5.2).

Regardless of their natural or artificial origin and nature, they have been classified as silicic, aluminic or a mixture of these two extreme chemical characters, i.e., silicic–aluminic or aluminic–silicic, depending, respectively, on whether the silicic or aluminic-chemical character prevails. In addition, when pozzolans on occasion also exhibit certain ferric characteristics, they are said to have a silicic–ferricaluminic chemical character or any other combination of those three characters (Rahhal and Talero 2010). According to these authors in some cases mineral additions that exhibit pozzolanic activity with lime [47–49] fail to do so with cement and vice versa (Mullick et al. 1986). According to ASTM C125 pozzolans must have a ($SiO_2 + Al_2O_3 + Fe_2O_3$) > 70%. Some authors (Al-Rawas et al. 2006) state that the pozzolanic activity increases with the increase of the total percentage of silicon dioxide plus aluminum oxide plus iron oxide.

Fly ash (*FA*): Some supplementary cementitious materials, like FA (a by-product from the coal-fired electricity production) have very slow hydration characteristics thus providing very little contribution to the early age strength (Mccartthy and Dhir 1999, 2004). Some authors (Roy 1987; Neville 1997) mention that high calcium FA (ASTM C type) is more reactive than (ASTM F type). FA is one the most used pozzolanic by-products and although the current replacement levels are below 40%, some authors showed that it is feasible to use more than 50% (Malhotra and Mehta 2005; Malhotra 2007; Van den Heede et al. 2010) as cement replacement.

Silica fume (*SF*): SF is a by-product from the production of the silicon metal with high pozzolanic activity. This by-product contributes for a denser concrete microstructure enhancing both strength and durability (Lachemi et al. 1998; Müller 2004; Song et al. 2010). Nevertheless, in some countries it is more expensive than Portland cement and its world production is limited to just 1 million tons (Khatib 2009).

Rice husk ash (*RHA*): RHA is a highly reactive pozzolan (Malhotra and Mehta 1996) obtained when rice husks are calcinated below the crystallization temperature at 780°C (Yu et al. 1999). RHA-based concrete has high strength and high durability performance (Anwar et al. 2000; Sousa Coutinho 2003; Zain et al. 2011). Since each ton of rice generates 40 kg of rice-husk ashes (Zerbino et al. 2011), this means that the annual world rice production of almost 600 million tons can generate almost 20 million tons of RHA. Usually after calcination the ashes are ground using a ball mill, however, Zerbino et al. (2011) mentioned that also non ground RHA can be used to replace 15% of Portland cement with similar mechanical and durability properties (Fig. 5.3). The use of ashes obtained from the calcination of other vegetable species as pozzolans in concrete have already been reported by several authors (Elinwa and Mahmood 2002; Elinwa and Ejeh 2004; Akram et al. 2009). But since these wastes are biodegradable attention must be directed to non organic-based pozzolans.

Sewage sludge ash (*SSA*): SSA is a siliceous material obtained by the calcination of water treatment wastes. Its pozzolanic activity depends on the chemical composition of the waste and on the calcination temperature (Monzo et al. 1999; Pan et al. 2002). The production of sewage sludge from waste water treatment plants is increasing all over the world. This kind of sludge includes the solid material left from sewage treatment processes. Specific sludge production in wastewater treatment varies widely from 35 g to 85 g dry solids per population equivalent per day (Davis 1996; Foladori et al. 2010). The total production of

Fig. 5.3 Natural rice husk ashes produced in Brazil. *Left* open field burning and aspect of the ashes; *right* storage and homogenization processes for laboratory studies (Zerbino et al. 2011)

sewage waste for the USA and the European Union (EU) approaches 17 Mt of dry solids/year (7 Mt in USA + 10 Mt in EU). Sewage sludge tends to accumulate heavy metals present in waste water and their concentration depends on the sludge origin (Fytili and Zabaniotou 2008).

One of the disposal solutions for this waste is through incineration, but this leads to hazardous emissions and even if new technologies are introduced for controlling emissions, almost 30% of sludge solids remains as ash (Malerious 2003). The incineration destroys the organic compounds, minimizes odours, greatly reduces sludge volume and has calorific value. Thus, the percentage of incinerated sludge is increasing all over the world (Lundin et al. 2004). The expected growth of world population and also the increase in the volume of waste water shows that sewage sludge ash will rise at a very fast pace in the subsequent years.

Waste ceramics and tungsten mine wastes: Several authors already confirmed the pozzolanic reactivity of ceramic wastes (Puertas et al. 2008; Naceri and Hamina 2009; Lavat et al. 2009) and more recently Pacheco-Torgal and Jalali et al. (2010) showed that concrete with 20% cement replacement by ceramic wastes, although it has minor strength loss has increased durability performance. Ceramic wastes can be separated into two categories in accordance with the source of raw materials. The first is, all fired wastes generated by structural ceramic factories that use only red pastes to manufacture their products, such as brick, blocks and roof tiles. The second is all fired wastes produced in stoneware ceramic such as wall, floor tiles and sanitary ware. These producers use red and white pastes, nevertheless, the usage of white paste is more frequent and much higher in volume. In Europe, the amount of wastes in the different production stages of the ceramic industry reaches some 3% to 7% of its global production, meaning millions of tons of calcined clays per year that can be used for Portland cement replacement. Some authors (Pacheco-Torgal et al. 2008a, 2009a; Choi et al. 2009) show that tungsten mine waste is an alumino-silicate source with ($SiO_2 + Al_2O_3 + Fe_2O_3$) > 70% with pozzolanic properties when submitted to a thermal treatment.

Recycled glass (*RG*): Finely ground waste glass having a particle size finer than 38 μm have pozzolanic behaviour and concrete containing ground glass exhibited a higher strength at both the early and late ages compared to fly ash concrete (Shao et al. 2000). Sahayan and Xu (2004) mention that fine glass powder could replace up to 30% of Portland cement. Other authors mention that alkali-silica reaction (ASR) and pozzolanic reactions observed for waste glass in concrete are very similar and suggest that they are closely related and may be simply various stages of each other (Federico and Chidiac 2009). Dyer and Dhir (2010) refer that high sodium content of the material raises concerns about whether the release of this element could ultimately exacerbate ASR and showed that powdered container glass is not suitable for controlling ASR.

Fluidized bed cracking catalyst (*FBCC*): Catalysts are widely used in the petrochemical industry and when the catalytic properties of this products are degraded, the deactivated catalyst must be replaced (Zornoza et al. 2009). FBCC, a waste from the petrochemical industry is a zeolite material containing more than 50% SiO_2 and about 40% Al_2O_3 (Pacewska et al. 2002; Bukowska et al. 2003). It improves concrete strength and increases its durability (Payá et al. 2009; Castellanos and Agredo 2010).

5.2.1.1 Tests for the Assessment of Pozzolanic Activity

These tests can be divided into two major groups, direct and indirect (Donatello et al. 2010b):

The direct test methods assess the reduction in $Ca(OH)_2$ with time as the pozzolanic reaction proceeds. These include XRD, TGA and chemical titration. The titrimetric methods include the Frattini test, the saturated lime test and the Chapelle test. The indirect tests measure a physical property related to the extent of the pozzolanic reaction. It includes the strength activity index (SAI), electrical conductivity and conduction calorimetry. Although SAI is very time-consuming when compared for instance with the conduction calorimetry that takes only 48 h (Rahhal 2002), some authors (Sinthaworn and Nimityongskul 2009) believe that this test is the most representative judgment strategy for pozzolanic activity evaluation, because not only does it show the contribution of pozzolanic reactivity,

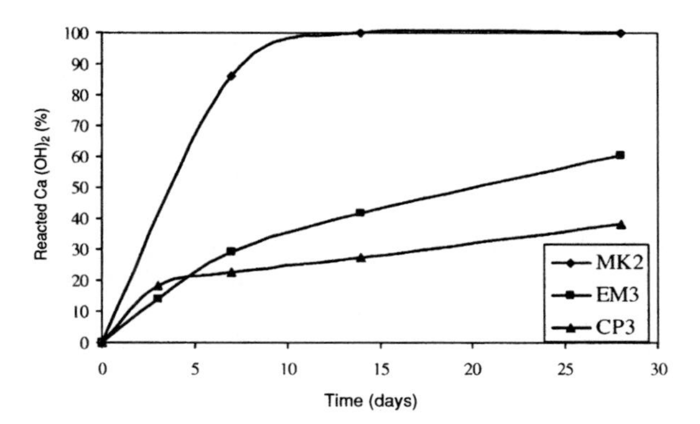

Fig. 5.4 Calcium hydroxide reaction versus curing time for metakaolin (MK2), earth of Milos (EM3) and ceramic powder (CP3)

but also includes the influence of packing effect in the improvement of the compressive strength. Infrared spectroscopy can also be used to assess the pozzolanic activity. The spectra are analyzed before and after treatments such as: acid–base attack and subsequent calcination at 1,000°C, calcination at 1,000°C and treatment with salicylic acid in a methanol medium (Luxán 1976). Wansom et al. (2010) confirm that spectroscopy can be used to evaluate the pozzolanic activity of RHA having a high correlation with SAI, only if the RHA have similar unburnt carbon content.

Moropoulou et al. (2004) compared the pozzolanic reactivity by DTA/TG analysis of two artificial pozzolans (metakaolin and ceramic powder) and a natural pozzolan (earth of Milos). The metakaolin has a higher $Ca(OH)_2$ consumption than the earth of Milos or the ceramic powder (Fig. 5.4).

The former has the lowest overall $Ca(OH)_2$ consumption although at 3 days curing, it has a higher consumption than earth of Milos. Some authors (Das and Yudhbir 2006) suggest the prediction of the pozzolanic activity of fly ash by strength index using a model based on the chemical composition and the fineness of the fly ash. The validity of the model increases with a fly ash replacement above 20%. Other authors (Franke and Sisomphon 2004; Sisomphon and Franke 2011) state that DTA analysis is not accurate for low calcium hydroxide content and suggest a chemical method to determine the calcium hydroxide content. Uzal et al. (2010) suggest that the compressive strength of lime-pozzolan 2.5 cm cubic specimens, pastes cured at 50°C is a more reliable indicator than the conventional pozzolanic tests. Gava and Prudêncio (2007a, b) present a comparison of the pozzolanic activity index results obtained from several test procedures prescripted by Brazilian, American and British Standards to evaluate three types of pozzolan: fly ash, silica fume and rice husk ash. These authors report that sometimes the different standards generate conflicting results. Pourkhorshidi et al. (2010a, b) mentioned that ASTM C618 SAI sometimes classifies as a high reactive pozzolan, some additions that lead to concrete with unsatisfactory performance. These authors mention that EN 196-5 had superior compatibility with real concrete performance. Donatello et al. (2010a) mentioned that the strength activity index of

ASTM C618 and of EN 196-5 is not suitable for evaluating the pozzolanic activity of incinerator sewage sludge ash. According to these authors since it is a low reactive pozzolan the 28-day curing period is not sufficient for it thus suggesting the Frattini test to evaluate its pozzolanic activity. Rahhal and Talero (2010) used the 2-day Frattini pozzolanicity test and the conduction calorimetry referring that the chemical composition of natural pozzolans, especially the reactive alumina content, is responsible for the contradictory (accelerating and retarding) effects of the rheology. Other authors (Donatello et al. 2010a) compared the pozzolanic activity of metakaolin, silica fume, coal fly ash, incinerated sewage sludge ash using the Frattini test, the saturated lime test and the strength activity index test. These authors mention that the saturated lime test is not a reliable test because it does not correlate with the other tests what could be explained by the different activator to pozzolan ratio. They recommend the use of the Frattini and SAI methods in combination with an independent determination of $Ca(OH)_2$ content (thermal or diffraction methods) for the assessment of the pozzolanic activity. The electrical conductivity method is also a very fast one. It was first suggested by Rassk and Bhaskar (1975) who measured the electrical conductivity of the amount of silica dissolved in a solution of hydrofluoric acid (HF) in which the pozzolan was dispersed. Luxan et al. (1989) suggested a rapid pozzolanic index given by the variation between the initial and final electrical conductivity of a calcium hydroxide–pozzolan suspension for a time of only 120 s. Feng et al. (2004) used the rapid method of Luxan to assess the pozzolanic properties of rice husk ash. These authors submitted the RHA to a hydrochloric acid pretreatment (1 N HCL) because acid leaching of the husk helps to obtain relative pure silica with high specific surface area. The results showed that the acid pretreatment enhanced the sensitivity of the pozzolanic activity of the RHA. Other authors (Tashiro et al. 1994) suggest that only 72 h testing is required to assess the pozzolanic behaviour with electrical conductivity if the lime-pastes were cured under steam at 70°C. McCarter and Tran (1996) also mentioned that the calcium hydroxide consumption after 72 h correlates with the electrical resistance of several artificial pozzolans. Paya et al. (2001) suggest a methodology based on the electrical conductivity that allows the assessment of the pozzolanic activity in less than 1 h. Although the test correlates with other standard methods it is not valid for high calcium fly ashes. Other authors (Villar-Cocina et al. 2003; Frias et al. 2005; Rosell-Lam et al. 2011) suggest a mathematic model based on the electrical conductivity measurements to describe the process in a kinetic or kinetic-diffusive regime. Dalinaidu et al. (2007) also describe a procedure to assess the pozzolanic activity by measuring the electrical conductivity of a pozzolan-lime solution. Other authors (Sinthaworn and Nimityongskul 2009) suggest that it is possible to assess the pozzolanic activity by measuring the change in electrical conductivity in a 28 h test. They studied pozzolans dispersed in an ordinary Portland cement using a temperature of 80°C. The results of the electrical conductivity show a high correlation with SAI.

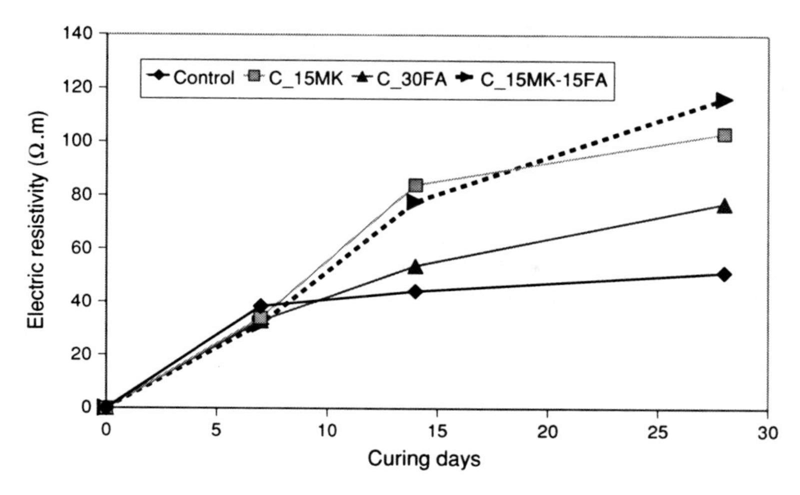

Fig. 5.5 Electric resistivity: C_15MK (mixture with 15% cement replacement with metakaolin; C_30F (mixture with 30% cement replacement with fly ash); C_15MK-15FA (mixture with 30% cement replacement with 15% metakaolin and 15% fly ash)

5.2.1.2 Influence of Pozzolans on Concrete Performance

The slow hydration characteristics of fly ash provides very little contribution to early age strength but for higher curing periods (118 days) fly ash concrete has a higher compressive strength than free fly ash concrete (Jiang et al. 2004). Concrete with organic ashes also show similar early strength loss (Elinwa and Mahmood 2002; Elinwa and Ejeh 2004; Akram et al. 2009). A simple way to compensate the early low reactivity of fly ash is by the use of a fly ash-metakaolin blend. Some authors (Pacheco-Torgal and Jalali 2011) showed that using a 15% fly ash and 15% metakaolin mixture, it is possible to obtain a compressive strength similar to the reference concrete. This mixture has the advantage of having a much lower electric resistivity (Fig. 5.5) and thus a lower corrosion risk.

The corrosion risk according to CEB 192 (Table 5.3), for the control mixture is always high. The mixtures with metakaolin perform in a similar way between the 7th and 14th curing days, meaning that the higher cement content is compensated by the synergetic effect between fly ash and metakaolin. From 14 days onward the mixture with 15% fly ash and 15% metakaolin shows a higher rising behaviour than the mixture with 15% cement replacement with metakaolin.

The electric resistivity of the fly ash/metakaolin mix is associated with a low risk corrosion after 3 weeks and seems to go to very low risk in a short period. Since electrical resistivity is one of the main parameters controlling the initiation and propagation of reinforcement corrosion (Ferreira and Jalali 2006; Ozkan and Gjorv 2008), the use of fly ash/metakaolin-based concrete seems to be a very effective option. Durability is in fact one of the key features associated with the use

Table 5.3 Corrosion risk according to concrete resistivity

Concrete resistivity (Ω m)	Corrosion risk
<50	Very high
50–100	High
100–200	Low
>200	Very low

of pozzolans in concrete (Malhotra 1978; Ramachandran 1932; Sabir et al. 2001. The high Portlandite consumption (Wild and Khatib 1997) reduces the matrix solubility being responsible for its increased durability. As a result , the resistance to acid attack increases and the corrosion risk decreases (Rodriguez-Camacho and Uribe-Afif 2002; Hossain et al. 2009; Fajardo and Valdez 2009; Kaid et al. 2009). The reduction of the calcium hydroxide also reduces the alkalinity helping to prevent the beginning of the alkali–silica reaction. Last but not the least, a high durability means less environmental impacts, for instance if we increase concrete durability from 50 to 500 years, we would reduce the environmental impacts by a factor of 10 (Mora 2007).

5.3 Concrete with Non Reactive Wastes

5.3.1 Construction and Demolition Wastes

Although the use of C&D wastes for the replacement of natural aggregates has been studied for almost 50 years (Malhotra 1978) today we still see many structures made with raw aggregates. The reasons for this are the low cost of raw aggregates, the lack of incentives, the use of low deposition costs and sometimes even the lack of technical regulations. For instance, only in 2006 Portugal approved a specific standard related to the production of concrete with recycled aggregates. Besides, the use of recycled aggregates in concrete is a scientific area with some gaps to be filled. Recycled aggregates manufactured in the laboratory are not contaminated with other wastes as it happens with aggregates obtained from C&Dw. Etxeberria et al. (2007) studied the performance of concrete with raw fine aggregates and different replacement percentages of coarse recycled aggregates referring that the use of a percentage of 25% is associated to a high compressive strength. It is noteworthy that these authors used a type I 52,5R cement which has a high amount of clinker and is not cost-efficient. Evangelista and Brito (2007) suggest that the use of fine recycled aggregates must not exceed 30%, otherwise the concrete performance could be at risk. Corinaldesi and Moriconi (2009) showed that mixture, it is possible to use 100% recycled aggregates without compressive strength loss as long as fly ash and silica fume are also used with a W/C = 0.4. Berndt (2009) also studied concretes with recycled aggregates, fly ash and blast furnace slag (W/C = 0.4) obtaining a compressive strength loss of just 2 MPa (Fig. 5.6).

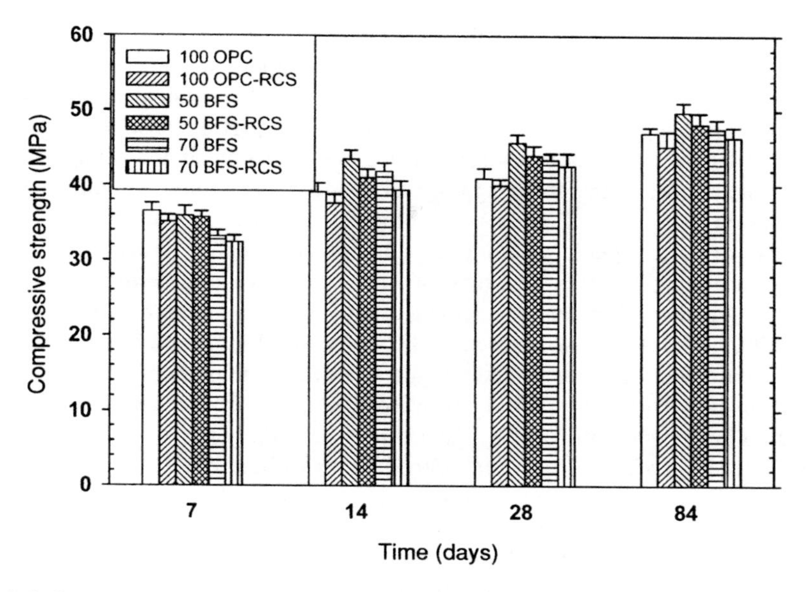

Fig. 5.6 Compressive strength of concrete with natural aggregates and small recycled aggregates (RCS) versus time

5.3.2 Vegetable Wastes

Several authors (Coatanlem et al. 2006) used pine wastes to produce lightweight concrete. The wood waste particles have a dimension between 5 and 10 mm and have been previously immersed in sodium silicate. This treatment increases the adhesion between the waste and the cement paste and also prevents attacks from insects or fungi. Branco et al. (2006) used cork wastes to replace natural aggregates observing a severe decrease in the compressive strength of concrete. A replacement between 10% and 30% leads to a compressive strength decrease between 27% and 53%.

5.3.3 Tyre Rubber Wastes

An estimated 1,000 million tyres reach the end of their useful lives every year (WBCSD 2010). At present enormous quantities of tyres are already stockpiled (whole tyre) or landfilled (shredded tyre), 3,000 millions inside EU and 1,000 millions in the USA (Oikonomou and Mavridou 2009). By the year 2030, the number of tyres from motor vehicles is expect to reach 1,200 million representing almost 5,000 millions tyres to be discarded on a regular basis. Tyre landfilling is responsible for serious ecological threats. Waste tyres disposal areas contribute to the reduction of biodiversity as they hold toxic and soluble components (Day et al. 1993). The implementation of the Lanfill Directive 1999/31/EC and the End of the

Life Vehicle Directive 2000/53/EC banned the landfill disposal of waste tyres creating the driving force behind the recycling of these wastes. Yet millions of tyres are just being buried all over the world. Tyre rubber wastes are already used for paving purposes; however, it can only recycle a part of these wastes (Vieira et al. 2010). In previous years several authors investigated the replacement of natural aggregates with rubber aggregates. Rubber aggregates are obtained from waste tyres using two different technologies: mechanical grinding at ambient temperature or cryogenic grinding at a temperature below the glass transition temperature (Nagdi 1993). The first method generates chipped rubber to replace coarse aggregates. The second method usually produces crumb rubber (Eleazer et al. 1992) to replace fine aggregates. Cairns et al. (2004) used long and angular coarse rubber aggregates with a maximum size of 20 mm obtaining concretes with an acceptable workability for a low rubber content. These authors reported a reduction in the workability for a higher rubber content, being that a rubber content of 50% led to a zero slump value. Other authors (Guneyisi et al. 2004) studied concretes containing silica fume, crumb rubber and tyre chips reporting a decrease in slump with increasing rubber content, being that a 50% rubber content leads to mixtures without any workability. Although investigations show that rubber aggregates lead to a decrease in concrete workability some authors reported no workability loss and others even observed the opposite behaviour; this means that workability is highly dependent on the characteristics of rubber aggregates. For the mechanical performance, Guneyisi et al. (2004) mentioned that the strength of concretes containing silica fume, crumb rubber and tyre chips decreases with the rubber content. These authors suggest that it is possible to produce a 40 MPa concrete replacing a volume of 15% of aggregates with rubber waste. Ghaly and Cahill (2005) studied the use of different percentages of rubber in concrete (5, 10 and 15%) by volume also noticing that as the rubber content increases it leads to a reduction in the compressive strength. Several authors mention the use of pre-treatments of rubber waste to increase the adhesion between the cement paste and the rubber, such as the use of a 10% NaOH saturated solution to wash the rubber surface for 20 min (Naik and Singh 1991; Naik et al. 1995). Raghavan et al. (1998) confirm that the immersion of rubber in an NaOH aqueous solution could improve the adhesion leading to a high strength performance of concrete rubber composites. The NaOH removes zinc stearate from the rubber surface, an additive responsible for the poor adhesion characteristics, enhancing the surface homogeneity (Segre et al. 2002). Segre and Joekes (2000) mention several pretreatments to improve the adhesion of rubber particles such as acid etching, plasma and the use of coupling agents. Cairns et al. (2004) suggested the use of rubber aggregates coated with a thin layer of cement paste (Fig. 5.7).

Albano et al. (2005) studied concrete composites containing scrap rubber previously treated with NaOH and silane in order to enhance the adhesion between the rubber and the cement paste without noticing significant changes, when compared to untreated rubber composites. Oiknomou et al. (2006) mentioned that the use of solid styrene butadiene (SBR) latex enhances the adherence between the rubber waste and the cement paste. Chou et al. (2010) suggest the pretreatment of

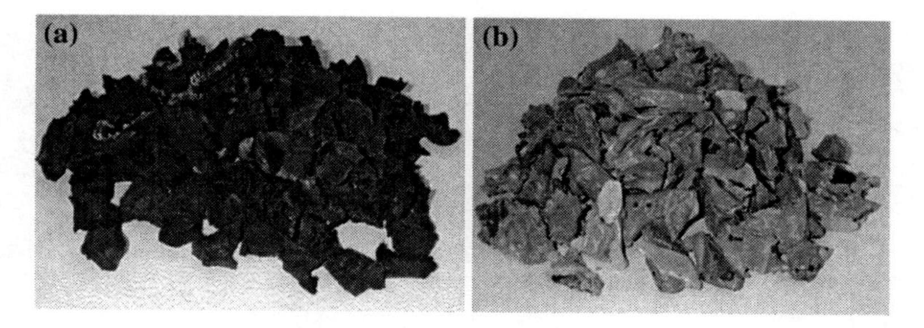

Fig. 5.7 A 20-mm rubber aggregate particles: **a** plain; **b** coated with cement paste

crumb rubber with organic sulfur stating that it can modify the rubber surface properties increasing the adhesion between the waste and the cement paste. Concrete composites containing tyre rubber waste are known for their high toughness (Li et al. 2004), having a high energy absorption capacity. Some authors (Balaha et al. 2007) report a 63.2% increase in the damping ratio (self-capacity to decrease the amplitude of free vibrations) for concrete containing 20% rubber particles. Other authors (Zheng et al. 2008a, b) confirmed the high damping potential of rubber waste concrete. They mention that concrete with ground rubber shows a 75.3% increase in the damping ratio and a 144% increase for crushed rubber concrete. This means that tyre waste concrete maybe especially recommended for concrete structures located in areas of severe earthquake risk and also for the production of railway sleepers. Ganjian et al. (2009) studied the durability of concrete containing scrap-tyre wastes assessed by water absorption and water permeability revealing that a percentage replacement of just 5% is associated with a more permeable concrete (36% increase) but not a more porous one. The durability of rubber waste concrete is a subject that still needs further investigations.

5.3.4 Polyethylene Terephthalate Wastes

Polyethylene terephthalate (PET) wastes represent one of the most common plastics in solid urban waste (Mello et al. 2009). In 2007 the world's annual consumption represented 250,000 million PET bottles (10 million tons of waste) with a growth annual increase of 15%. In the US, 50,000 million bottles are landfilled each year (Gore 2009). Since PET waste is not biodegradable it can remain in nature for hundreds of years. The workability of PET wastes based concrete is influenced by the fact that PET wastes were previously submitted to a pretreatment (Choi et al. 2005; Batayneh et al. 2007). It remains to be seen whether these pretreatments have an environmental impact that shadows the ecological benefits of reusing PET wastes in concrete. Kim et al. (2008) mentioned that the use of a volume of just 0.25% PET fibres can reduce the plastic shrinkage, increasing PET fibres' volume beyond 0.25% does very little for the shrinkage

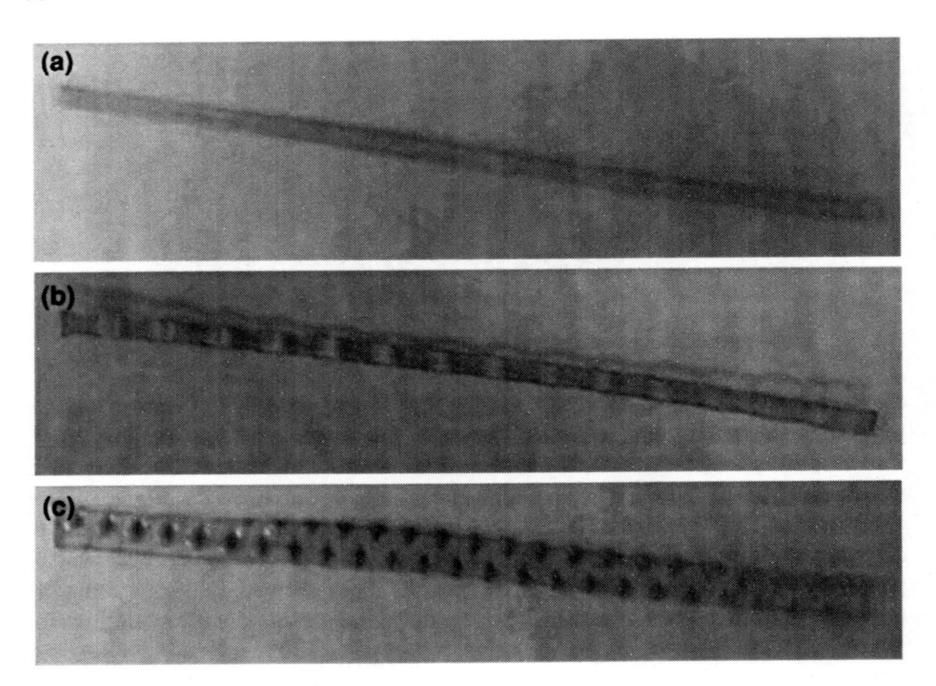

Fig. 5.8 Geometry of recycled 50-mm long PET fibres: **a** straight type (cross section 0.5 × 1 mm); **b** crimped type (cross section 0.3 × 1.2 mm); **c** embossed type (cross section 0.2 × 1.3 mm)

reduction. Kim et al. (2010) confirm the concrete crack control ability of PET fibre composites. These fibres are obtained from melted PET waste to form a roll-type sheet. Then the sheet is cut into 0.5 mm long fibres and a deforming machine is used to change the fibre surface geometry (Fig. 5.8). They mentioned that the use of a volume of just 0.25% PET fibres can reduce the plastic shrinkage, increasing the PET fibre volume beyond 0.25% does very little for the shrinkage reduction. The results confirmed that the embossed type fibre, the one that has the best mechanical resistance, leads to the best shrinkage performance. Since investigations on concrete shrinkage performance use pretreated PET fibre, these investigations should study which pretreatment has the lowest environmental impact.

Choi et al. (2005) mention that the replacement of fine aggregates with pretreated PET/GBFS aggregates (5–15 mm) leads to minor decreases in the compressive strength. This result is related to the spherical and smooth shape of the aggregates used in the investigation which are made of PET waste and ground granulated blast furnace slag (GBFS). According to these authors these aggregates are made inside a mixer with an inner temperature of 250 ± 10°C. At this temperature the PET particles start to melt down and then they mix with the GBFS particles resulting in a composite aggregate with a PET core and a GBFS surface (Fig. 5.9).

For a 25% PET/GBFS aggregates replacement, the mixtures with a W/C = 0.45 and 3 curing days lost just 6.4% of the compressive strength. After 28 curing

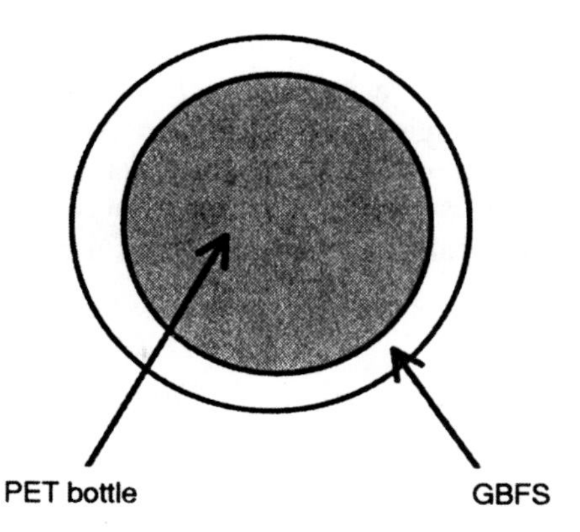

Fig. 5.9 PET waste/GBFS aggregates

days the compressive strength loss reaches just 9.1%. Increasing the replacement percentage increases the compressive strength loss but not in a proportional manner, for instance a 75% replacement in the mixtures with a W/B = 0.45 and 3 curing days lost just 16.5% of the compressive strength. This means that these pretreated PET/GBFS aggregates perform in almost a similar way as natural aggregates do. However, other authors (Batayneh et al. 2007) mentioned a severe compressive strength decrease of 72% for just 20% volume replacement of untreated PET waste. This behaviour is very different from the one reported by previous authors which means that using untreated PET waste implies the use of a very small volume in order to obtain an acceptable compressive strength concrete. Marzouk et al. (2007) found that the tensile and compressive strengths of crushed PET wastes concrete composites have a similar loss pattern. Ochi et al. (2007) mention that the use of pretreated 30 mm long indented PET fibres can lead to a tensile strength increase for volume replacements of up to 1.5%. Silva et al. (2005) mention that recycled monofilament PET fibres increases the toughness indexes of cementitious composites. Also, Hannawi et al. (2010) mention that PET and polycarbonate waste composites have a high energy absorbing behaviour even with a high waste content. Yet, investigations that can clarify which pretreatment have the lowest environmental impact and maximizes the toughness characteristics of PET waste concrete are needed. Regarding the durability characteristics of PET fibres, is worth mention that they degrade in the alkaline environment of the cement paste. The mechanism of PET degradation involves a depolymerization reaction that breaks the polymer chain splitting it into two groups (the aromatical ring and the aliphatic ester). Several other polymeric wastes have been investigated about their potential to be used as aggregate replacement in cementitious composites. Laukaitis et al. (2005) studied the development of lightweight thermo-insulating cementitious composites containing crumbled polystyrene waste and spherical blown polystyrene waste. Other authors (Panyakapo and Panyakapo 2008) studied concretes with ground thermosetting polymer (melamine) waste reporting a reduction in the compressive strength with a waste content increase, which is related to the poor adhesion between the waste plastic and the cement paste. Yadav (2008) also confirms the strength reduction associated with concrete polymeric waste composites. Nevertheless, the results show that in spite of the compressive strength reduction it is possible to produce non-loading-bearing lightweight concrete. Studies on the durability of other polymeric wastes are scarce but some authors (Wang et al. 1987; Houget 1992) already report that polyester and acrylic fibres can suffer from chemical degradation when immersed in the alkaline environment of a cement paste.

5.4 Concrete with Organic Polymers

The use of polymers in concrete goes back to 1923 when for the first time a patent was issued for a concrete floor with natural latex, being that Portland cement was used only as filler. It was only in 1924 that the first patent on hydraulic binders modified with polymers (Ohama 1998) was issued. However, only in the 1950s appeared the first uses of concrete modified with polymers, particularly in the rehabilitation of concrete structures. At present, three kinds of polymer-based concretes can be separated due to their different natures. One group is related to polymer modified concrete (PMC) or polymer cement concrete (PCC) and is composed of aggregates and a binder matrix where phases generated by the hydration of Portland cement coexists with polymeric phases. Another group is related to polymer impregnated concrete (PIC), in which concrete is impregnated with a monomer of low viscosity, usually of methyl methacrylate in order to fill its porous structure. A third group is related to polymer concrete (PC), this group is composed of aggregates and a polymer matrix without Portland cement (Fowler 1999). For PMC, additives are added to concrete during the mixing stage, usually in the form of a colloidal suspension of latex, powder or as water-soluble polymers or liquids, and the literature usually refers to more uses of the polymer of styrene-butadiene (SBR) of polyacrylic-ester (PAE), polyethylene vinyl acetate (EVA). Figure 5.10 presents a conclusive model to explain the hydration of cement with polymers.

Step 1: Immediately after mixing the cement and the polymer particles remain dispersed in water;

Step 2: Some polymer particles are deposed in the surface of cement and aggregates;

Step 3: Depending on the curing conditions polymer can coalescence into a film;

Step 4: The hydration of the cement proceeds and polymer film formation increases.

Concrete with polymers has superior durability over ordinary Portland cement concrete, assessed by resistance to acid attack (Monteny et al. 2001), resistance to freeze-thaw (Chmielewska 2007), resistance to diffusion of chlorides (Yang et al. 2008). The explanations for this difference in behaviour are due to, a lower porosity of the formation of a polymer film inside the pores (Rossignolo 2005) and to a low permeability to water

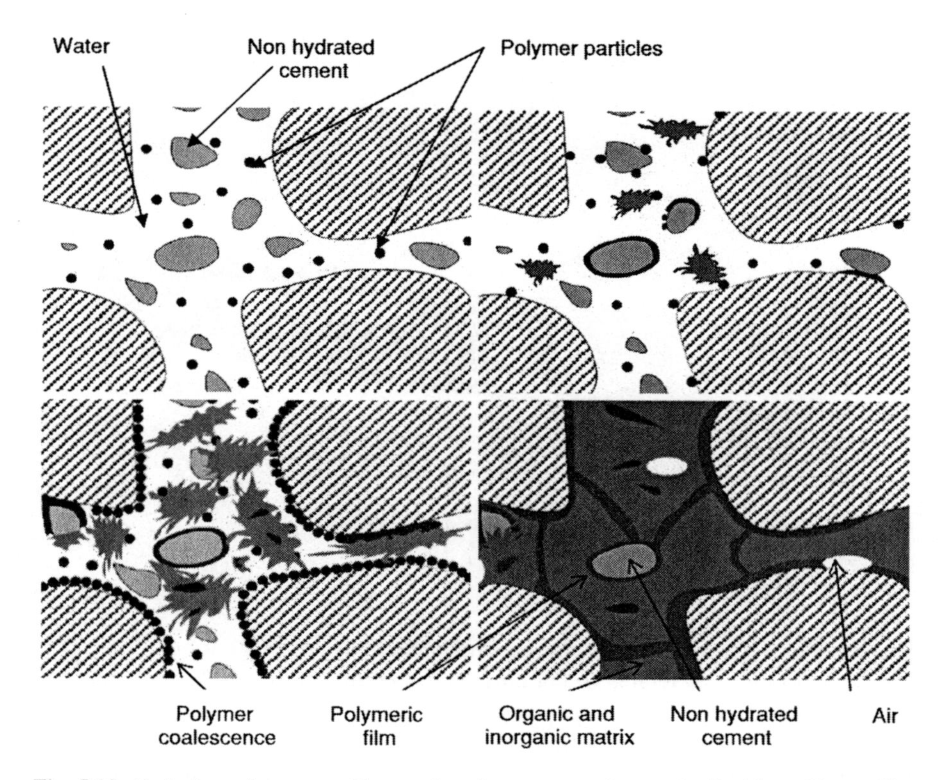

Fig. 5.10 Hydration of cement with organic polymers according to the Beeldens–Ohama–Van Gemert model (Van Gemert et al. 2005)

access (Neelamegan et al. 2007). Several authors (Ogawa et al. 2007; Shirai et al. 2007) show that polymer impregnation of concrete materials may lead to an increase in durability depending of the type of polymers used. More recently, Pacheco-Torgal and Jalali (2009) showed that concrete with polymer impregnation performs better than concrete with polymer addition not only in terms of the chemical resistance but also in terms of cost-efficiency. Sustainable trends related to the use of organic polymers include the development of repair materials for retrofitting of concrete structures (Ohama 2010; Van Gemert and Knapen 2010).

5.5 Self-Sensing Concrete

Ordinary Portland cement (OPC) concrete structures deterioration is a very common phenomenon and the number of premature cases OPC structures disintegration is overwhelming. Beyond the durability problems originated by

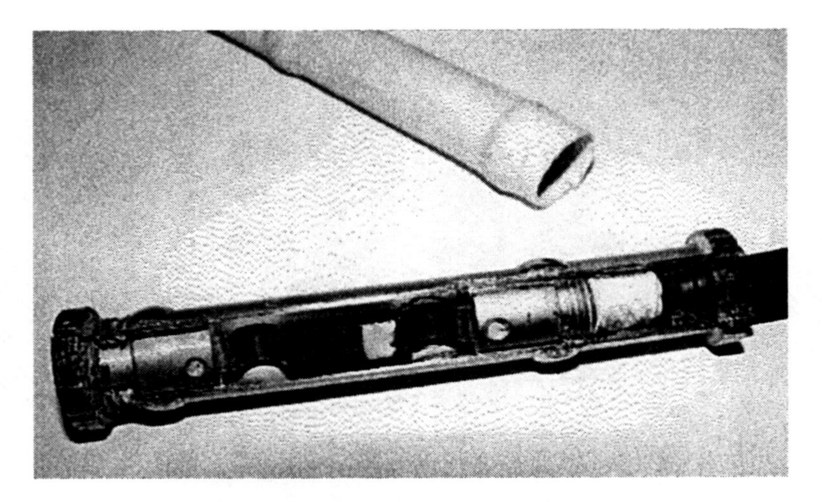

Fig. 5.11 Device for concrete deformation assessment

imperfect concrete placement and curing operations, the real issue about OPC concrete durability is related to the intrinsic properties of this material. It presents a higher permeability that allows water and other aggressive elements to enter, leading to carbonation and chloride ion attack resulting in corrosion problems. The lack of regular inspections in order to assess the conservation state of OPC concrete structures worsens the problem and contributes to premature degradation leading to possible OPC concrete structures failure with the inevitable loss of human lives. Currently, the monitoring of OPC concrete structures requires the use of extensometers bonded to its outer surface or embedded within the concrete (Fig. 5.11).

This procedure is expensive, and in the case of embedded devices may be responsible for property loss and induce premature concrete degradation. Sensing is a fundamental aspect of a smart structure. A smart structure must have the ability to sense stimuli and to be able to respond to it in an appropriate manner. Structural composites which are themselves sensors are multifunctional materials. Carbon fibre cement–matrix composite materials are gaining momentum due to the reduction of carbon fiber cost and also to the sensing performance of carbon fibre reinforced concrete-based structures. The sensing ability of carbon fibres reinforced concrete is due to the electric conductivity provide by the carbon fibres. Research on fibre reinforced concrete (FRC) composites shows that these new concretes have increased tensile and flexural strengths, lower drying shrinkage and improved freeze–thaw durability (Park and Lee 1993; Pigeon et al. 1996a, b; Banthia and Sheng 1996). Cement paste is electrically conductive with a DC resistivity at 28th day curing around 5,000 Ω m at room temperature. The addition of short (5 mm) carbon fibres (0.5% by weight of cement) decreases the resistivity of carbon fibre concrete to just 200 Ω m in the presence of silica fume which provides fibre dispersion (Chung 2002). The resistivity of concrete reinforced with carbon fibres is influenced by the volume and the size of carbon fibres, and also by the saturation degree of the cement–matrix (Wen and Chung 2001; Chen et al. 2004). For carbon fibre concrete electrical resistance increases with traction stress and decreases upon compression (piezoresistivite property). The explanation for that behaviour is related to the fact that tension leads to micro crack opening, thus increasing concrete resistivity. Therefore, carbon fibre concrete can act as a self-monitoring strain sensor (Wen and Chung 2005). The applications for piezoresistive behaviour of carbon fibre concrete include weighing, traffic monitoring, building facility management, building security and structural vibration control. It also enables building facility management through the use of the occupancy of each room to control lighting, heating, cooling and ventilation. Carbon fibre concrete can also be used to assess its own damage enabling structural health monitoring (Reza et al. 2003; Wen and Chung 2006). This ability can also be used to assess damage evolution (Cao and Chung 2002), this property will enable real-time monitoring, which is a crucial tool to avoid structure failure. Since electric resistivity decreases with rising temperature this means that reinforced carbon fibre concrete can act as a thermistor, thus providing information about thermal control, structural operation control or even about hazard monitoring. Another smart application for this new material is related to its use as a heating element thus improving freeze–thaw durability by heating concrete and preventing freezing. Therefore, deicing (removal of ice) and anti-icing (prevention of icing) are no longer needed. Moreover, it can also be used in heating of building floors. Since carbon fibre reinforced concrete can act as an electric conductor it also provides lightning protection. Due to the wave reflection of this new material it can also be used for electromagnetic interference (EMI) shielding, thus enabling the construction of cell-phone proof buildings, a very useful property for detention facilities and probably class rooms and conference halls. Concrete composites with 1.5% carbon fibre volume enable a shielding effect of 40 dB at 1 Ghz (Gnecco 1999). Chung (2001) showed that due to small diameter and skin effect carbon nanofibres can be more effective than regular carbon fibres for EMI shielding. A fundamental application of smart concrete composites relates to corrosion control of steel reinforcing rebars using cathodic protection. This method requires the application of an electric current that forces electrons to go towards the steel rebar, thereby making the steel a cathode. Concrete itself has low electric conductivity but using carbon fibre reinforced concrete facilitates cathodic protection. Concrete with carbon fibre and silica fume reduces the driving voltage required for cathodic protection in 28% compared to concrete without carbon fibres (Chung 2000). Another way to address steel corrosion when using carbon fibre concrete composites is by assessing the presence of chloride ions using electric resistivity and then applying an electric current that forces the removal of chloride ions away from steel rebars.

5.6 Concretes Based on New Binders

5.6.1 Sulfo-Aluminate Cement

Sulfo-aluminate cements are also called belitic cements or belite sulfo-aluminate cements (Sharp et al. 1999). These cements have been widely produced in China as "Third Cement Series". They contain more than 50% of belite (Ca_2SiO_4) while Portland cements contains only between 15% to 30%, having instead a major phase of alite (Ca_3SiO_5) between 50% to 70% (Morsli et al. 2007). Sulfo-aluminate cements use lower calcination temperatures than Portland cement, around 1,200°C, and also use limestone with less carbonate content, thus representing a 10% reduction in CO_2 emissions (Quillin 2001), however, recent investigations mention a reduction of carbon dioxide emissions of up to 35% (Martín-Sedeño et al. 2010). Being low heat cements, they have the advantage to be used in massive applications such as large dams, however, as a drawback they have low early strength (Cuberos et al. 2009) due to the low hydration of the belite phase. Regarding their grindability, Winnefeld and Lothenbach (2010) state this type of clinker is easier to grind when compared to Portland cement. Popescu et al. (2003) describe a mixture easier to grind than Portland cement (high ferro-belite clinker) but also mention a mixture (sulphoferroaluminate-belite clinker) with high resistance to milling. According to Scrivener and Kirkpatrick (2008) other disadvantages arise due to the difficulties of finding cheap alumina sources and to the fact that the use of a high content of sulfate may lead to increased acid rain.

5.6.2 Magnesium Phosphate Cement

According to Yang and Wu (1999) the first research on commercial magnesium phosphate was published in 1984. They are also called magnesium phosphate ceramics and belong to the group of chemically bonded ceramics (Buj et al. 2009). Magnesium phosphate cement paste is prepared by mixing dead-burned magnesia powder and potassium di-hydrogen phosphate with water. Then an exothermic reaction occurs, due to the dissolution of magnesia and phosphate ions which form hydrates that gather around struvite ($NH_4MgPO_46H_2O$) nuclei (Soudée and Péra 2000). These binders show high-early strength and low shrinkage, however, and depending on the reactivity of the magnesia a fast setting can take place making the mixture unsuitable for any use (Qiao et al. 2009). Qiao et al. (2010) mentioned the setting time as well as the mechanical properties of this binder are influenced by the magnesia/phosphate (M/P) molar ratio, thus recommending a M/P = 8. This author states that magnesium phosphate cement mortar can be used for the repair of Portland cement concrete structures. Magnesium phosphate cement shows high durability (Yang et al. 2000, 2002). Other authors confirm this but mention that the alkali resistance of the magnesium phosphate cement is poor (Wang et al. 2011).

5.6.3 Alkali-Activated Binders

These binders are synthesised from aluminosilicate raw materials with alkaline solutions and are associated with lower carbon dioxide emissions than OPC. Alkali-activated binders show high mechanical performance and also superior acid and abrasion resistance than OPC concretes. These new binders can use ashes from power stations or mining and quarrying wastes as raw materials and can be used for the immobilization of radioactive and toxic wastes which gives them an undeniable environmental value (Van Deventer et al. 2010). The carbon dioxide emissions of alkali-activated binders is a subject of some controversy. Joseph Davidovits (Davidovits et al. 1990, Davidovits 1999) was the first author to address the carbon dioxide emissions of these binders stating that they generate just 0.184 tons of CO_2 per ton of binder (Table 5.4).

Duxson et al. (2007) do not confirm these numbers, they mention that although the CO_2 emissions generated during the production of Na_2O are very high, yet the production of alkali-activated binders is associated to a level of carbon dioxide emissions lower than the emissions generated in the production of OPC. According to these authors the reductions can go from 50% to 100% (Fig. 5.12). The results show how the activator weight influences the carbon dioxide emissions of the alkali-activated binders. Lower reductions in CO_2 emissions are obtained when using raw materials that require a heat treatment such as metakaolin. The higher reductions are obtained with fly ash. The calculation include 72 kg CO_2/ton related to the metakaolin production.

According to these authors the emission reductions should be regarded as conservative because the Na_2O is a by-product generated in the production of chlorine, a material used by the plastic industry in the production of PVC and other materials, hence the CO_2 emissions can be partially allocated to that industry. Duxson and Van Deventer (2009) mention an independent study made by Zeobond Pty LtD in which a low emissions Portland cement (0.67 ton./ton.) and alkali-activated binders were compared, reporting that the latter had 80% lower CO_2 emissions. Weil et al. (2009) mentioned that the sodium hydroxide and the sodium silicate are the responsibles for the majority of CO_2 emissions in alkali-activated binders. These authors compared OPC concrete and alkali-activated concrete with similar durability reporting that the latter have 70% lower CO_2 emissions which confirm the aforementioned reductions.

5.6.3.1 Historical Background

The development of alkali-activated binders had a major contribution in the 1940s with the work of Purdon (1940). The author used blast furnace slag activated with sodium hydroxide. According to him, the process was developed in two steps. During the first step, liberation of silica, aluminum and calcium hydroxide took place. After that, the formation of silica and aluminum hydrates would happen as

Table 5.4 Emissions of CO_2 during the production of 1 ton of alkali-activated binder (Davidovits 1999)

Constituent	Calcination conditions	Ton. CO_2/ton of constituent	CO_2/ton of alkali-activated binder
SA07	800°C	0.17	0.095
Metakaolin	750°C	0.15	0.035
GBFS	–	–	–
Potassium silicate	1,200°C	0.30	0.034
Grinding energy	–	–	0.020
Total of emissions per ton of alkali-activated binder			0.184

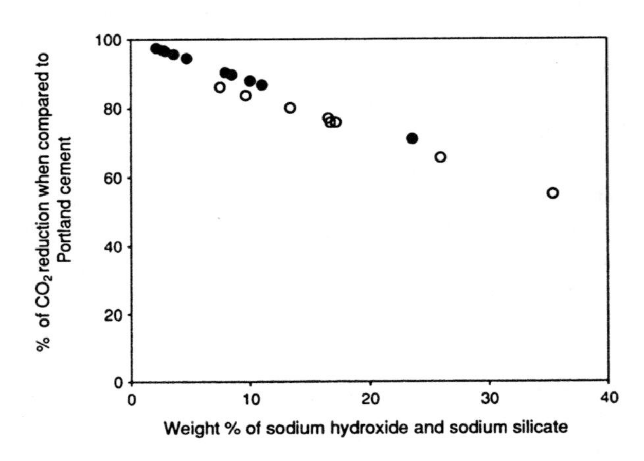

Fig. 5.12 CO_2 emissions as a function of the alkaline solution and the raw material (solid points-fly ash, open circles-metakaolin)

well as the regeneration of the alkali solution. However, it was Glukhovsky (1959) who first investigated the binders used in ancient Roman and Egyptian constructions. He concluded that they were composed of aluminosilicate calcium hydrates similar to those of portland cement and crystalline phases of analcite, a natural rock that would explain the durability of those binders. Based on his investigations Glukhovsky developed a new type of binder that he named "soil–cement", the word soil is used because it seemed like a ground rock and the word cement due to its cementitious capacity. The "soil–cement" was obtained from ground aluminosilicate mixed with rich alkali industrial wastes. A large part of the investigations about alkali-activated binders is related to the activation of blast furnace slag, known as "Alkali-slag cement" or "Alkali-activated slag cement". Blast furnace slag is a by-product of the iron production industry, having a high content of calcium which is due to the use of calcium carbonate in the calcination operations. Being a low performance cementitious material, it can achieve high compression strength when an alkaline activator is used. Shi and Day (1995) mention that the alkali-activation with $Na_2O.nSiO_3$ led to a compression strength of 160 MPa after 90 days curing at room temperature. However, Glukhovsky et al. (1980) had already made crucial investigations about the activation of blast furnace slag: (a) identifying the hydration products as being composed of calcium silicate hydrates and calcium and sodium aluminosilicate hydrates, (b) noticing that clay minerals when submitted to alkali-activation formed aluminium silicate hydrates (zeolite). This author (1981) classified the alkaline activators into six groups, where M is an alkali ion:

1. Alkalis, MOH
2. Weak acid salts, M_2CO_3, M_2SO_3, M_3PO_4, MF
3. Silicates, $M_2O.nSiO_3$
4. Aluminates, $M_2O.nAl_2O_3$
5. Aluminosilicates, $M_2O.Al_2O_3{\cdot}(2\text{–}6)SiO_2$
6. Strong acid salts, M_2SO_4

Investigations in the field of alkali-activation had an exponential increase after the research results of Davidovits (1979) who developed and patented binders obtained from the alkali-activation of metakaolin, having named it after the term "geopolymer" in 1978. For the chemical designation of the geopolymer Davidovits suggests the name "polysialates", in which Sialate is an abbreviation for aluminosilicate oxide. The sialate network is composed of tetrahedral anions $[SiO_4]^{4-}$ and $[AlO_4]^{5-}$ sharing the oxygen, which needs positive ions such as (Na^+, K^+, Li^+, Ca^{++}, Na^+, Ba^{++}, $NH_4{}^+$, H_3O^+) to compensate the electric charge of Al^{3+} in tetrahedral coordination (after dehydroxilation the aluminium changes from coordination 6 (octahedral) to coordination 4 (tetrahedral). The Polysialate has the following empiric formulae:

$$M_n\{-(SiO_2)_z - AlO_2\}_n,\ w\,H_2O$$

in which: n is the degree of polymerization, z is 1, 2 or 3, and M is an alkali cation, such as potassium or sodium, generating different types of Poly(sialates) (Fig. 5.13). According to Davidovits, geopolymers are polymers because they transform, polymerize and harden at low temperature. But they are also geopolymers, because they are inorganic, hard and stable at high temperature and also non inflammable.

5.6.3.2 Prime Materials

Types

Theoretically, any material composed of silica and aluminium can be alkali-activated. The investigations performed so far have used the following prime materials:

1. Kaolinitic clays (Davidovits and Sawyer 1985; Davidovits, 1989; Rahier et al. 1996; Rahier et al. 1997; Barbosa et al. 2000);

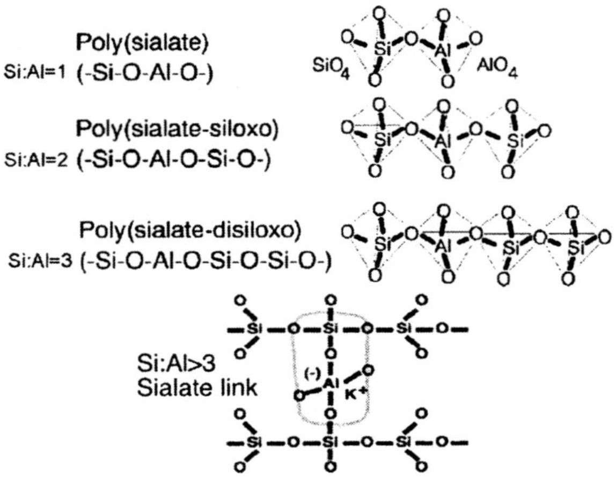

Fig. 5.13 Poly(sialates) structures (Davidovits 2005)

Poly(sialate)
Si:Al=1 (-Si-O-Al-O-)

Poly(sialate-siloxo)
Si:Al=2 (-Si-O-Al-O-Si-O-)

Poly(sialate-disiloxo)
Si:Al=3 (-Si-O-Al-O-Si-O-Si-O-)

Si:Al>3
Sialate link

2. Metakaolin (Davidovits 1999; Barbosa et al. 2000; Alonso and Palomo 2001a,b; Pinto 2004);
3. Fly ashes (Palomo et al. 1999a; Fernandez-Jimenez and Palomo 2005;
4. Blast furnace slag (Purdon 1940; Wang Shao-Dong and Scrivener 1995; Fernandez-Jimenez et al. 1999; Adolf and Bazan 2007);
5. Mixtures of fly ashes and slag (Puertas et al. 2000; Puertas and Fernandez-Jimenez 2003);
6. Mixtures of fly ashes and metakaolin (Swanepoel and Strydom 2002);
7. Mixture of slag and metakaolin (Cheng and Chiu 2003);
8. Mixture of slag and red mud (Zhihua et al. 2002, 2003);
9. Mixtures of fly ashes and non calcined materials like kaolin and stilbite (Xu and Van Deventer 2002).

Xu and Van Deventer (2000) studied the alkali-activation of 16 natural alumino-silicate minerals, having reported that all of them show some reactivity. Stilbite activated with potassium hydroxide displayed the highest mechanical strength. For Gourley 2003, the use of metakaolin has the advantage of being white, although it needs several hours of thermal treatment lowering its economic competitiveness.

Fernandez-Jimenez and Palomo (2003) studied the reactivity of fly ashes to be alkali-activated, obtaining some reactivity for all of them. The most important reactivity parameters were the reactive silica content, the amorphous phase content, the gradation and the calcium content, gradation and calcium content. These authors claim that iron and calcium do not influence the mechanical strength because they are not found in the main reaction products. This claim is opposite to the results of Van Jaarsveld et al. (2003) according to whom fly ashes in the presence of calcium lead to higher mechanical strengths. The thermal treatment of alumino-silicate materials give rise to changes in their structure with an increase in the amorphous phase. For that kind of structural change XRD analysis is not appropriate, hence those changes are currently assessed by infrared emission spectra analysis (FTIR). According to some authors (Lee and Van Deventer 2002a) the FTIR is accurate to detect small structural changes, being the most appropriate technique to assess structural changes of amorphous alumino-silicate materials. Therefore, the mechanical strength of alkali-activated binders depends on the structural conditions of the alumino-silicate materials, being that natural materials lead to lower mechanical performance. Higher mechanical strengths are associated with materials submitted to calcinations such as fly ashes, blast furnace slag and metakaolin. As it happens in pozzolanic reactivity, alkali-activation reactivity depends on the amorphous content of silica and aluminium. The reactivity is linked to the material structure, being higher for higher amorphous contents.

5.6.3.3 Blaine Fineness

Andersson and Gram (1988) found that the mechanical strengths of alkali-activated specimens do not increase very much when Blaine fineness of blast furnace slag increase from 5,300 to 6,700 cm^2/g. Talling and Brandstetr (1989) obtained an optimum strength for a slag Blaine fineness of 4,000 cm^2/g. As for Wang et al. (1994) the optimum Blaine fineness depends on the type of slag and varies between 4,000 and 5,500 cm^2/g. Granizo (1998) studied alkali-activated metakaolin claiming that Blaine fineness is the most important parameter. Other authors (Fernandez-Jimenez et al. 1999) studied alkali-activated slag mortars obtaining lower mechanical strength when the Blaine fineness increase from 4,500 to 9,000 cm^2/g. They also conclude that among several parameters Blaine fineness was the least relevant one. This results are opposite to the ones noticed by Brough and Atkinson (2002). These authors studied alkali-activated slag, reporting that an increase in the Blaine fineness from 3,320 to 5,500 cm^2/g leads to an increase in mechanical strength from 65 to 100 MPa. Other authors (Weng et al. 2005) reported that increasing metakaolin Blaine fineness from 15,670 to 25,550 m^2/g lead to an increase in mechanical strength from 55 to 74 MPa. According to them, this raises the available amount of aluminium to react with the alkaline activator. Therefore, more aluminium means more $[Al(OH)_4]^-$ tetrahedral groups able to attract negatively charged groups and so increasing the reacted species amount. Generally speaking, blast furnace slag with higher Blaine fineness means more reactivity, but at the same time more hydration water, leading to higher porosity and lower strength.

5.6.3.4 Alkaline Activators

The most used alkaline activators are a mixture of sodium or potassium hydroxide (NaOH, KOH) with sodium waterglass ($n SiO_2 Na_2O$) or potassium waterglass ($n SiO_2 K_2O$) (Granizo 1998; Davidovits 1999; Fernandez-Jimenez et al. 1999; Palomo et al. 1999b; Barbosa et al. 2000; Bakharev et al. 2003; Escalante-Garcia et al. 2002; Swanepoel and Strydom 2002; Xu and Van Deventer 2002; Hardjito et al. 2002b). Katz (1998) studied alkali-activated slags reporting an increase in mechanical strength when the concentration of the activator increases. Other authors have reported the same behaviour using alkali-activated metakaolin (Wang et al. 2005; Pinto 2004). However, for the alkali-activation of fly ashes Palomo et al. (1999b) reported that an activator with a 12 M concentration leads to better results than using a 18 M concentration. Some authors used free waterglass alkaline activators, having noticed lower mechanical performances (Palomo et al. 1999b; Pinto 2004). For Palomo et al. (1999b) the alkaline activator plays a crucial role in the polymerization reaction, behaving more swiftly when the soluble silica are present. This statement is also shared by Criado et al. (2005), according to whom waterglass favours the polymerization process leading to a reaction product with more Si and more mechanical strength. Van Jaarsveld et al. (1997), Van Jaarsveld and Van Deventer (1999) claim that the H_2O/SiO_2 molar ratio is very important in the study of alkali-activated mixtures. However,this statement is not confirmed by others (Hardjito et al. 2002a), who claim that it does not influence the mechanical strength. Kirschner and Harmuth (2002) studied the activation of metakaolin with NaOH and waterglass having noticed that mechanical strength increases when Na_2O/SiO_2 molar ratio decreases. Other authors (Rowles and O'Connor 2003) also studied the alkali-activation of metakaolin, noticing that the mechanical strength was higher for a molar ratio Si/Al/Na of 2.5:1:1.3. According to Fernandez-Jimenez et al. (1999) the most relevant parameters that influence the mechanical strength of alkali-activated blast furnace slag mortars are: the nature of the alkaline activator, the concentration of the activator, the curing temperature, and the least one was Blaine fineness. These authors also noticed that the optimum concentration of the alkaline activator varies from 3% to 5% of Na_2O of slag mass. Using a Na_2O amount above those limits, gives rise to cost inefficient mixtures with efflorescence problems. Other authors (Bakharev et al. 1999) also claim that the activation of blast furnace slag with a waterglass-based activator (Ms = 1.25) leads to the highest mechanical strength. However, these authors achieved the optimum strength using 8% of Na_2O. Xu and Van Deventer (2000) confirm that the use of waterglass increases the dissolution of the prime materials. They studied the alkali-activation of different natural alumino-silicate minerals, having noticed that the majority of them could not provide enough Si to start the geopolymerization, thus needing additional soluble silica. Puertas et al. (2000) studied pastes of fly ashes/slag and claim that the compressive strength is influenced by the concentration of the sodium hydroxide. They also reported a compressive strength increase with a slag content increase, explaining it with the higher reactivity of the blast furnace slag. Lee and Van Deventer (2002b) reported increased dissolution due to an excess of alkali, but it also give rise to a formation of a aluminosilicate gel in the first curing ages leading to a mechanical strength decrease. Some authors noticed that the waterglass/sodium hydroxide molar ratio influence the compressive strength (Pinto 2004; Hardjito et al. 2002a), as the use of a molar ratio of 2.5 leads to an expressive strength increase of alkali-activated fly ash (Hardjito et al. 2002a). Krizan and Zivanovic (2002) studied the alkali-activation of blast furnace slag with waterglass and metasilicate noticing that the highest strength was obtained for Ms (1.2–1.5). Other authors (Xie and Xi 2001) used fly ash-based mixtures activated with waterglass and sodium hydroxide. They report that when the NaOH amount increases lowering the sílica modulus (Ms = 1.64) the silicate excess is crystallized being responsible for a higher compressive strength. Fernandez-Jimenez and Palomo (2005) studied the activation of fly ash, with several activators in which the Na_2O content changed from 5% to 15%. They concluded that the SiO_2/Na_2O molar ratio as well as the W/B ratio influence the mechanical strength. They noticed that using an Na_2O content of 5.5% by fly ash mass leads to a very low pH affecting the reaction development in a negative way and also that the Na_2O content increase leads to a mechanical strength increase, as the use of 14% of Na_2O by fly ash mass leads to an optimum mechanical performance.

5.6.3.5 Use of Additives

Although Davidovits investigations used alumino-silicate materials calcium free, the fact is that Pyrament cement is composed of 80% of portland cement (allegedly due to its low cost) and also 20% of geopolymeric materials activated by potassium carbonate having citric acid as a retarder (Davidovits 1994a). Also the patented geopolymeric cements PZ-Geopoly® and Geopolycem® have 11% of CaO in his composition (Davidovits 1994b) confirming the importance of calcium in alkali-activated binders. Alonso and Palomo (2001a) studied alkali-activated metakaolin/calcium hydroxide-based mixtures reporting that an increase of the amount of metakaolin over calcium hydroxide gives rise to an increase of the formation of alkaline alumino-silicate compounds due to the increase in alumino-silicate species dissolved and also that temperature increase hastened the reaction, lowering ion mobility. In another investigation the authors (Alonso and Palomo 2001b) studied the influence of the sodium hydroxide concentration on the nature of the reaction products formed, concluding that this parameter has a crucial role on them:

- When the alkaline activator concentration is 10 M or higher, dissolution of $Ca(OH)_2$ is very difficult due to the presence of hydroxides (OH^-) meaning that there will not be enough formation of CSH gel, instead sodium-based aluminosilicate is formed. When that occurs it attracts OH^- to its structure lowering their amount and allowing the formation of CSH gel as a secondary reaction product.
- When the alkaline activator concentration is lower than 5 M, the amount of hydroxides (OH^-) is very low so the dissolution of calcium hydroxide takes place meaning that enough Ca^{2+} will be present to form CSH gel. Besides, low alkaline concentration media prevents the metakaolin dissolution so that there is not enough dissolved aluminium for the formation of alkaline alumino-silicates, meaning that silica will be free to form CSH. Lee and Van Deventer (2002b) studied the influence of inorganic salt in alkali-activated mixtures of fly ash and

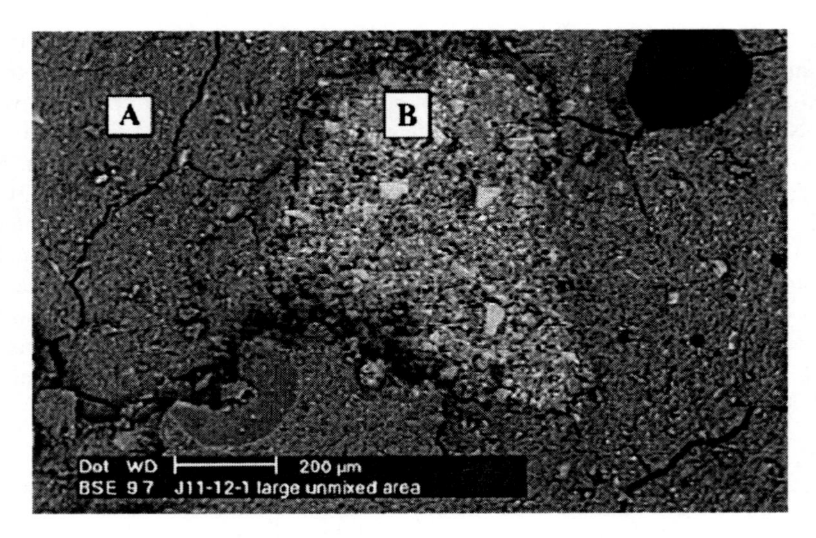

Fig. 5.14 Formation of geopolimeric gel (**a**) and CSH gel (**b**) (Yip and Van Deventer 2003)

kaolin having reported that strength and durability are negatively affected by chloride salts, because they formed crystals inside the structure lowering their mechanical strength. They also found that carbonate salts are beneficial because they diminish the amount of dissolved water preventing the hydrolytic attack. Other authors (Yip and Van Deventer 2003) reported that calcium increases the mechanical strength due to the formation of Ca–Al–Si amorphous structures. They found the coexistence of geopolymeric gel and CSH (Fig. 5.14) and suggested that the formation of those two phases would explain the durability of ancient binders.

Escalante-Garcia et al. (2003) studied alkali-activated slag mortars with 10% replacement with geothermal waste silica and noticed that when the activator is made just of NaOH the replacement always gives rise to a mechanical strength increase. However, when the activator uses waterglass the strength increase happens only before 7 days curing, after that time the replacement leads to lower strength due to an excess of silica that leads to the formation of H_2SiO_3.

5.6.3.6 Constituents Mixing Order

Some authors believe that the optimum mixing order for alkali-activated binders is the following. First , the solids are mixed, apart from that the activator is prepared and put to rest and finally the activator is mixed with the solids (Van Jaarsveld et al. 1998; Swanepoel and Strydom 2002; Cheng and Chiu 2003). Pinto (2004) studied the alkali-activation of metakaolin confirming that this mixing order leads to the best results. Cheng and Chiu (2003) reported the best mixing order as: mix metakaolin and potassium hydroxide for 10 min, then add waterglass and slag and

Fig. 5.15 Mixing the activator with the solid components

mix for 5 min, place the mixtures in 50 × 50 × 50 mm^3 molds and vibrate them for more 5 min. Sumajow and Rangan (2006) report the mixture of solid components for 3 min in a vertical mixer followed by the introduction of the activator (silicate and hydroxide), and a new mixing period of 4 min (Fig. 5.15).

5.6.3.7 Curing Conditions

Different raw materials using different activators were associated with different curing conditions. Pinto (2004) mentioned that the cure in a saturated environment reduces the mechanical performance of alkali-activated binders and that specimens with a mass ratio activator/metakaolin = 0.75, lost 6% of the mixing water by evaporation and needed to be covered to prevent this. Brough and Atkinson (2002) studied the activation of blast furnace slag with sodium silicate stating that the cure at 80°C for 12 h allowed a strength gain from 7 to 72 MPa when compared to a curing at 20°C. These results are very different from the ones reported by Wang et al. (1994). These authors used mortars based on alkali-activated blast furnace slag noticing that the temperature depends on the type of activator used. However, the influence is irrelevant when high Blaine fineness GBFS is used or when the activator has a high alcalinity. Kirschner and Harmuth (2002) report a maximum strength for a cure at 75°C during 4 h. They confirm that curing alkali-activated binders in water leads to a decrease in the mechanical performance. According to Criado et al. (2005) the activation of fly ash for certain curing conditions leads to a carbonation process that lowers the pH and reduces the mechanical performance. For Bakharev (2005) the temperature is a crucial factor in the alkaline activation of fly ash due to the activation barrier that must be overcome in order to start the reaction process. This is confirmed by the work of Katz (1998) who observed a dramatic increase in the mechanical performance for increasing curing temperature. The activation energy is higher for fly ashes than for GBFS hence the curing temperature is more important for the former. These results are confirmed by other authors (Fernandez-Jimenez and Puertas 1997; Fernandez-Jimenez et al. 1999; Puertas et al. 2000). Sumajow and Rangan (2006) mention a new variable known as "resting time" which corresponds to the time between the placement of the alkali-activated binder into the moulds and the time when this moulds were cured with increase temperature.

5.6.3.8 Mechanical Strength

For the same water/binder ratio some authors reported that alkali-activated binders present a higher mechanical strength than OPC binders. Wang (1991) reports a case of an alkali-activated slag concrete with a 125 MPa compressive strength. Davidovits (1994a) reports having obtained a 20 MPa compressive strength just after 4 h increasing to 70–100 MPa after 28 days curing. Other authors (Bakharev et al. 1999) used slag pastes activated with NaOH and waterglass (w/b = 0.5) having reported lower but rising mechanical strength, respectively 8, 16 and 39 MPa for 1day, 7days and 28 days curing. However, when slag mortars were used (aggregate/slag = 2) a rise in the earlier mechanical strength was noticed, respectively 9, 21 and 26 MPa. Fernandez-Jimenez et al. (1999) studied slag mortars (aggregate/slag = 2 and W/B = 0.51) activated with NaOH and waterglass reporting 100 MPa for compressive strength and 11 MPa for tensile strength. Zhihua et al. (2003) used alkali-activated slag/red mud mixtures (Ms = 1.2) also reporting high strength performance, respectively $fc^{1day} = 20$ MPa; $fc^{28days} = 56$ MPa; $ft^{1day} = 3.3$; $ft^{28days} = 8.4$ MPa. For metakaolin-based mortars activated with NaOH and waterglass Pinto (2004) reported for compressive strength, 53 and 60 MPa, respectively for 7 and 28 days curing. When alkali-activated metakaolin concrete was used he reported 71 and 77 MPa for compressive strength at 7 and 28 days curing, and also 10 MPa for tensile strength. Fernandez-Jimenez and Palomo (2005) used slag/fly ashes mixtures activated with NaOH and waterglass (W/B = 0.35) reporting a 90 MPa compressive strength just after 20 h. Bakharev (2005) studied fly ash pastes activated with NaOH and waterglass (W/B = 0.3) reporting a 60 MPa compressive strength just after 2 days.

5.6.3.9 Durability

Resistance to acid attack: Several authors reported that chemical resistance is one of the major advantages of alkali-activated binders over OPC. Davidovits et al. (1990) reported mass losses of 6% and 7% for alkali-activated binders immersed in 5% concentration hydrochloric and sulfuric acids during 4 weeks. For the same conditions he also reported that OPC concretes suffered mass losses between 78% and 95% (Fig. 5.16). Allahverdi and Skvára (2001a, b, 2005, 2007) studied alkali-activated binders based on fly ash and blast furnace slag reporting that their acid resistance depends on the type of acid and also on the pH of the acid solution.

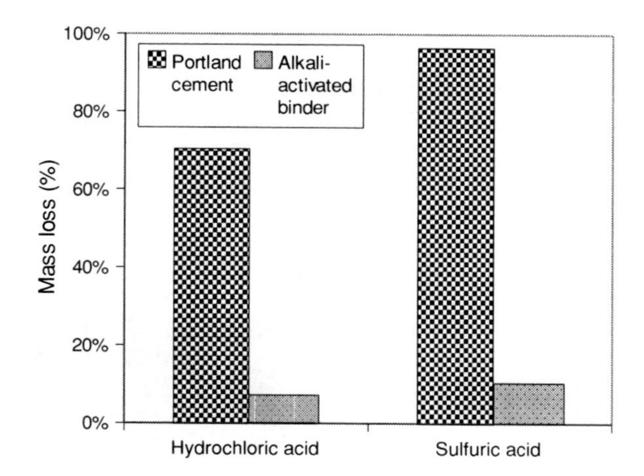

Fig. 5.16 Resistance to acid attack of Portland cement and alkali-activated binders

The same authors observed that the solutions of nitric acid with a pH less than 2, led to faster degradation of these binders. For acid solutions with a pH higher than 3, the corrosion mechanism is similar for the two acids. Bakharev et al. (2003) studied OPC concretes and slag concretes activated with NaOH and waterglass, immersed in an acetic acid solution (pH = 4) during one year. They reported a 33% strength loss for the former and 47% for OPC concretes. They claim that the strength loss is influenced by Ca content, being that OPC concretes had a Ca content of 64% and alkali-activated slag concretes just 39%. Besides, slag compounds have lower Ca/Si molar ratio and are more stable in acid médium.

As for OPC concrete calcium compounds, they possess high Ca/Si molar ratios and react with acetic acid forming acetic calcium compounds, which is very soluble. Song et al. (2005) also confirm that alkali-activated fly ash concretes have high chemical resistance, when immersed in a 10% concentration sulfuric acid solution during 8 weeks, they showed mass and strength losses of 3% and 35%. Gourley and Johnson (2005) mentioned that a OPC concrete with a service life of 50 years lose 25% of its mass after 80 immersions cycles in a sulfuric acid solution (pH = 1) while an alkali-activated concrete required 1,400 immersions cycles to lose the same mass, thus meaning a service life of 900 years.

Resistance to high temperatures: Concretes based on Portland cement show a weak performance when subjected to a thermal phase and when the temperature rises above 300°C, they begin to desintegrate. As for alkali-activated binders they show a high stability when submitted to high temperatures even around 1,000°C (Pawlasova and Skavara 2007). Bortnovsky et al. (2007) studied the alkali-activation of metakaolin and shale wastes reporting a high mechanical performance after a thermal phase (Table 5.5).

The specimens show some slight strength loss between 600°C and 1,000°C, however in some cases they show a strength increase at 1,200°C. Kong et al. (2008) studied alkali-activated metakaolin binders observing that the residual strength after a thermal phase up to 800°C is influenced by the Si/Al ratio. The

Table 5.5 Percentage of residual strength of alkali-activated binders after being submitted to high temperatures

Mix	Residual flexural strength (%)			Residual compressive strength (%)		
	600 °C	900 °C	1,200 °C	600 °C	900 °C	1,200 °C
H160	90	90	157	81	82	110
H110	93	93	145	88	76	122
K80	–	91	155	–	76	85

Fig. 5.17 Fire resistance testing of alkali-activated plates

higher residual strength was obtained by the mixtures with a Si/Al ratio between 1.5 and 1.7.

Resistance to fire: Krivenko and Guziy (2007) found that alkali-activated binders show a high performance in the resistance to fire (Fig. 5.17), thus suggesting that this material is suitable for use in works for which the fire has serious consequences like it happens in tunnels and tall buildings.

Perná et al. (2007) confirmed that alkali-activated binders can be used as a 120 min anti-fire material in accordance with related standards of the Czech Republic. The anti-fire material must show a temperature lower than 120°C in the opposite side of the fire action.

Resistance to freeze–thaw cycles: Dolezal et al. (2007) reported the loss of only 30% of the resistance in alkali-activated fly ash binders after being subjected to 150 freeze–thaw cycles. Other authors (Bortnovsky et al. 2007) studied the resistance of alkali-activated slag-waste shales based binders reporting a high compressive strength even after 100 freeze–thaw cycles (Table 5.6).

Abrasion resistance: Pacheco-Torgal (2007) studied alkali-activated mine waste binders reporting a high resistance to abrasion in the Los Angeles test machine (Fig. 5.18).

As a comparison OPC concrete specimens of two different strength classes (C20/25 and C30/37) had a much lower performance. The higher abrasion resistance was achieved in alkali-activated mine waste paste specimens. This result is

Table 5.6 Compressive strength after freeze-thaw cycles

Mix	Compressive strength (MPa)	Compressive strength after 50 cycles (MPa)	Compressive strength after 100 cycles (MPa)
K80	91	82	75
H110	105	84	90
K125	88	79	89
H160	110	85	79

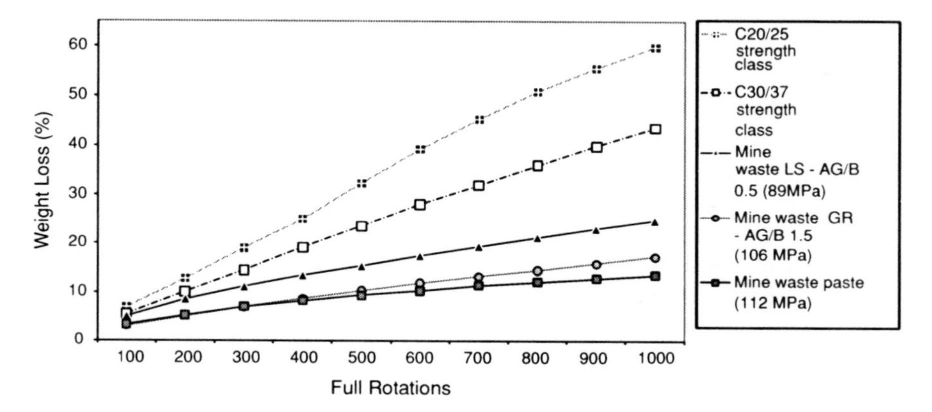

Fig. 5.18 Abrasion resistance when using the Los Angeles test machine for OPC concrete and alkali-activated mine waste binders

partially related to the fact that mine waste paste had the highest compressive strength. After 1,000 full rotations the worst alkali-activated mix (LS) had not even a 30% loss of its mass. Hu et al. (2008) also confirm the high abrasion resistance of alkali-activated binders.

5.6.3.10 Reuse of Mine Wastes

Pacheco-Torgal et al. (2007b) study the reuse of mine waste mud from the tugsten mine Panasqueira as raw material in alkali-activated binders. Panasqueira is an underground mine utilising room and pillar mining methods situated in central Portugal on the southern edge of the Sierra da Estrela mountain range, a natural park, near the Sierra do Açor, a protected landscape near the Zezere river. Tungsten and tin have been mined in the Panasqueira mine since 1890. In the mid-1980s, Panasqueira had over 1,000 employees in its underground mining and plant operations that processed 600,000 tons of ore per annum to produce in excess of 2,000 tons of tungsten oxide (WO_3) just about of 0.3% of excavated rock. During the mining process two types of mine wastes are generated, coarse aggregates derived from rock blastings and waste mud conveyed by pipelines into lagoons

Fig. 5.19 Panasqueira tungsten mine wastes. Coarse aggregates and waste mud lagoon (Pacheco-Torgal et al. 2007a)

amounting for several million tons and still generating almost 100 tons/day. Figure 5.19 shows the environmental impact of Panasqueira mine wastes.

The mineralogical composition of mine waste mud deduced from an XRD study using a Rigaku Geigerflex diffractometer, was muscovite and quartz, which were identified by their characteristic patterns. The chemical composition of the mine waste mud obtained by atomic absorption using a spectrophotometer shows that they consist essentially of silica and alumina, contaminated with arsenic and sulfur and with a high content of iron and potassium oxide. The mine waste was thermally treated to achieve the dehydroxylated state.The XRD patterns indicate that the dehydroxylation did not result in the collapse of the muscovite structure. Peak area measurements revealed that about 12% of muscovite survived calcination at 950°C. Molecular changes during dehydroxylation were also examined with infrared emission spectra (FTIR), confirming a decrease in the absorption peaks at 3,600–3,700 (OH stretch), however the main muscovite peak did not disappear totally indicating only a partial transformation. Compressive strength of alkali-activated waste mortars was used to evaluate the dehydroxylation degree, $50 \times 50 \times 50\ mm^3$ mortar cubes were cast to study the compressive strength and their development with curing time. Calcination of mine waste at 950°C during 120 min considerably increases the compressive strength at 28 days curing. This is due to the structural dehydroxylation process which leads to an amorphous product of dehydration muscovite. This result is consistent with XRD results. Compressive strength data related to alkali-activated mortars made with, respectively raw waste mud and calcined waste mud showed an increase of more than 400% justifying the thermal treatment (Pacheco-Torgal et al. 2005). The studies about the influence of the mix design on the strength of the new binder show that the highest strength is associated with a mixture of sodium hydroxide and sodium silicate with a mass ratio of 1:2.5 (Pacheco-Torgal et al. 2008b). The replacement of mine waste with 10% of calcium hydroxide was found to lead to the highest compressive strengths. Alkali-activated mine waste binders show a compressive strength higher than 30 MPa after only one day, reaching almost 70 MPa after 28 days curing and 90 MPa at 90 days curing. These binders also show a flexural strength exceeding

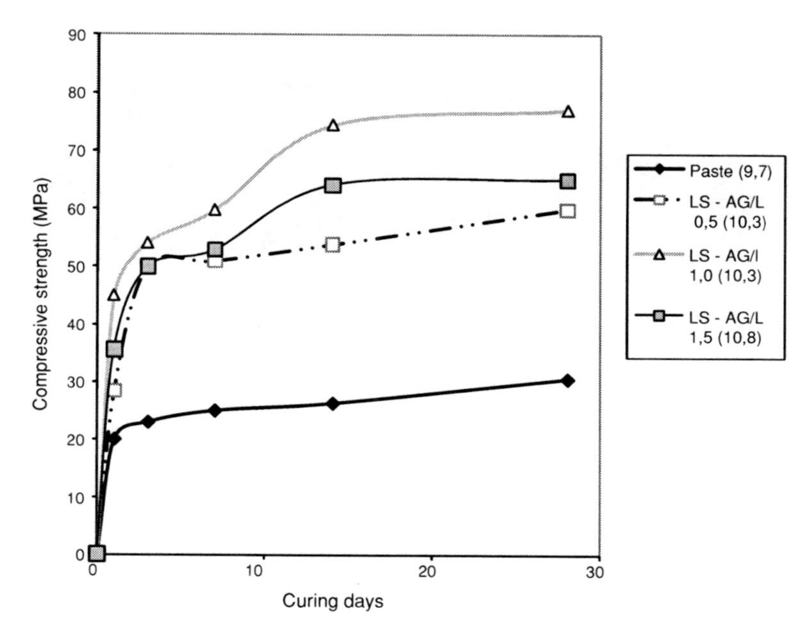

Fig. 5.20 Compressive strength according to aggregate/binder mass ratio and H_2O/Na_2O molar ratio in mine waste binders (Torgal et al. 2007)

10 MPa (Pacheco-Torgal et al. 2008b). It was also found that the aggregate type influences the compressive strength of the alkali-activated mine waste binders. This behavior is rather different from Portland cement in which the compressive strength depends crucially on the cement hydration. One suggests that strength behaviour can only be explained by the chemically active role played by the aggregates, which may be due to the dissolution of quartz in the presence of alkalis enhancing the bond strength between paste and aggregates. This is confirmed by the compressive strength of alkali-activated mine waste paste which is significantly lower than the compressive strength of the mortars (Fig. 5.20).

Regarding water absorption by immersion (2–3%) the alkali-activated mine waste binders have a much lower absorption than Portland cement concrete (more than 10%). As for the static modulus of elasticity, the range values between 29 MPa and 34 GPa (Pacheco-Torgal et al. 2008b) are not much different from those currently associated with Portland cement concretes. The new binders show a high abrasion resistance (mass loss below 25%) while Portland cement concrete shows severe mass loss between 40% and 60%. Mine waste binders show a high resistance to acid attack (Fig. 5.21). The mixture with the best performance shows an average mass loss of just 2.6% after being submitted to the attack of (sulfuric, hydrochloric and nitric) acids for 28 days, while the mass loss for Portland cement concretes is more than twice that value (Pacheco-Torgal et al. 2007a).

Fig. 5.21 Cubic specimens after immersion in a solution of nitric acid during 24 h (Pacheco-Torgal et al. 2007a)

Alkali-activated mine waste binders show an outstanding adhesion to Portland cement concretes which is almost independent of the concrete substrate surface treatment. As a result they can be used in retrofitting operations of concrete structures with the advantage of being much more cost-effective than the current pre-pack commercial mortars (Pacheco-Torgal et al. 2006; 2008c). The use of this new binder as a building material requires the assessment of its environmental performance. For that, leaching tests have been carried out according to DIN 38414-S4. Leaching results show that all chemical parameters are below the limits established by the standard and can be considered as inert materials. The reuse of mine wastes as raw materials in alkali-activated binders will reduce the environmental impact of their deposition, reduce the consumption of non renewable resources and simultaneously reduce the emissions of GHGs (Pacheco-Torgal et al. 2009b).

5.6.3.11 Toxic Metals Immobilization

The alkali-activation of aluminosilicate materials has been used as a way to immobilize toxic metals. Van Jaarsveld et al. (1997) used alkali-activated binders to immobilize mine waste toxic metals reporting the following reductions in the leaching concentration Fe(99%), Cu(99%), As(95%), they also reported less efficiency immobilization for Ti(65%), Ni(40%), Zn(40%). Other authors (Van Jaarsveld and Van Deventer 1999) used alkali-activated fly ash/metakaolin mixtures with minor quantities of copper and lead, reporting that the oxidation state of the ion (by the time when leaching tests are carried out) influences the ionic radius and thus the leaching degree. They note that the immobilization of lead (Pb) and copper (Cu) happens not only by an encapsulation mechanism but also due to a chemical bonding between the ion and the chemical structure of the binder. They report that the presence of lead and copper influences the properties of the hardened binder and also that higher radius ions are better immobilized. According to Hermann et al. (1999) the use of alkali-activated binders is a good way to immobilize a wide range of harmful constituents such as toxic metals, hydrocarbonates and even nuclear wastes in a final product with high durability and costing

Fig. 5.22 Immobilization of toxic wastes: **a** alkali-activated waste fresh mix; **b** big bags with hardened binder; **c** placement of the big bags in a landfill (Hermann et al. 1999)

Table 5.7 Some ionic radius

Ion	Cu^{2+}	Fe^{3+}	Cd^{2+}	As^{5+}	Pb^{2+}	Hg^{2+}
Raio (Å)	0.72	0.64	0.97	0.46	1.20	1.10

much less than the current vitrification process. These authors describe an immobilization procedure using big bags containing alkali-activated binders and that will later be placed in a waste landfill (Fig. 5.22).

Table 5.7 shows different ionic radius in different ions, thus confirming that some ions like arsenic will have more tendency to be leached . The results of Davidovits et al. (1990) about the immobilization of mine toxic metals in alkali-activated mixtures, confirm the existence of a relationship between toxic metals immobilization and ionic radius (Table 5.8).

Palomo and Palacios (2003) studied the capabilities of alkali-activated fly ash mixtures to immobilize chromium and lead, reporting that lead immobilization is achieved with a low reduction in the mechanical strength of the binder. Nevertheless, they were not able to immobilize chromium because that metal disturbs the geopolymerization mechanism, due to the formation of $Na_2OCrO_4{\cdot}4H_2O$ which is a very soluble compound. Other authors (Qian et al. 2003) studied the immobilization of mercuric ions Hg^{2+} in alkali-activated slag pastes. Claiming that the immobilization depends on a mechanical encapsulation due to the low permeability of these binders and also to a chemical mechanism in which Hg^{2+} replaces Ca^{+} in the CSH structure. These authors also reported a similar beaviour in slag

Table 5.8 Heavy metal efficiency for alkali-activated wastes (Davidovits et al. 1990)

Constituent	Cu	Cd	Pb	Mo	Cr	Zn	Ni	V
Efficiency (%)	98	85	60	60	50	40	15	12
Ionic radius (Å)	0.72/0.96	0.97/1.14	1.20	0.93	0.63/0.89	0.74	0.69	0.59

pastes containing zinc ions (Qian et al. 2003a). Phair et al. (2004) used alkali-activated fly ash binders reporting higher immobilization efficiency with lead ions (Pb) than with copper ions (Cu), because the former gives rise to smaller compounds. They also reported that Pb and Cu immobilization involves an encapsulation mechanism as well as the formation of an insoluble phase with silica and aluminium. Other authors (Bankowski et al. 2004) used metakaolin-based mixtures activated with NaOH and waterglass. These authors report concentration reductions for arsenic, strontium, selenium and barium immobilization. They also reported that the concentration of chromium, copper and molybdenum had not changed and that nickel, vanadium and zinc had increased their concentration. Vinsova et al. (2007) refer that alkali-activated binders show a good performance in the immobilization of lead, cadmium and chromium, being less effective for the immobilization of arsenic. Lancellotti et al. (2010) show that alkali-activated metakaolin binders are able to immobilize toxic metals present in fly ash due to the incineration of municipal solid wastes.

5.7 Conclusions

For its volume and economic importance in the construction sector, exacerbated by future projections, or for the environmental impacts associated with it, this chapter on binders and concrete is the longest chapter of this book. Portland cement production represents the majority of the total CO_2 emissions of concrete, hence the use of pozzolans as cement replacement can allow major carbon dioxide reductions and also increase the service life of concrete structures; furthermore, in the case of waste pozzolans it also reduces the need for disposal areas. Regarding the replacement of natural aggregates with recycled aggregates one can see that the technical regulations are far behind the state of the art of the investigations made in this field. Investigations on the use of C&DW for the replacement of natural aggregates has been studied for almost 50 years and more recently even on aggregates obtained from several kinds of wastes (wood, plastic or rubber). However, not even the former case has become a routine situation in the construction industry, which is due to the fact that natural aggregates have extremely competitive prices and also to the low taxes used for the landfill of C&DW. As for organic polymer-based concretes, they are associated to an increased durability but as a downside, they are heavily dependent on the petroleum industry. This chapter has also addressed the case of sensing concrete. Although these materials are still under investigation,in the future they will allow to build structures with important functions, such as those concerning the control of their own deformation and degradation leading to a high service life. The last part of this chapter was dedicated to new binders alternative to Portland cement. Sulfo-aluminate cements use lower calcination temperatures than Portland cement, thus generating less carbon dioxide emissions, but since cheap sources of raw alumina are not available it seems that this binder cannot hinder the monopoly of Portland cement. Magnesium phosphate cement mortar could be used as a repair material for current concrete structures, however, questions related to its exothermic reaction and its fast setting must be addressed in future investigations Alkali-activated binders emit less carbon dioxide than Portland cement, have a high mechanical strength, a high resistance to abrasion and acid attack and they can be used for the immobilization of toxic metals. However, it is not expected that in a short term alkali-activated concrete will replace Portland cement concrete, it is more likely that this new binder could be used in the retrofitting of current concrete structures.

References

Adolf Z, Bazan J (2007) Utilisation of metallurgical slags as raw material basis for preparation of alkali activated materials. In: Proceedings of the 2007–Alkali Activated Materials–Research, Production and Utilization 3rd Conference. Agentura Action, Prague, Czech Republic 11–19

Agarwal SK (2006) Pozzolanic activity of various siliceous materials. Cem Concr Res 36:1735–739. doi:10.1016/j.cemconres.2004.06.025

Akram T, Memon S, Obaid H (2009) Production of low cost self compacting concrete using bagasse ash. Constr Build Mater 23:703–712. doi:10.1016/j.conbuildmat.2008.02.012

Albano C, Camacho N, Reyes J, Feliu J, Hernández M (2005) Influence of scrap rubber addition to Portland concrete composites: destructive and non-destructive testing. Compos Struct 71:439–446. doi:10.1016/j.compstruct.2005.09.037

Allahverdi A, Skvara F (2007) Evaluating the potential application of fly ash/blast furnace slag geopolymer material for inhibiting acid corrosion, a comparative study. Alkali activated materials-research, production and utilization 3rd conference, Prague, Czech Republic, pp 21–37

Allahverdi A, Škvára F (2001a) Nitric acid attack on hardened paste of geopolymeric cements-Part 1. Ceram-Silikaty 45:81–88. http://www.geopolymery.eu/aitom/upload/documents/publikace/2001/2001_03_081.pdf

Allahverdi A, Škvára F (2001b) Nitric acid attack on hardened paste of geopolymeric cements-Part 2. Ceram-Silikaty 45:143–149. http://www.ceramics-silikaty.cz/2001/2001_04_143.htm

Allahverdi A, Škvára F (2005) sulfuric acid attack on hardened paste of geopolymer cements—Part 1. Mechanism of corrosion at relatively high concentrations. Ceram-Silikaty 49:225–229. http://www.geopolymery.eu/aitom/upload/documents/publikace/2005/allahverdi_2005_04_2251.pdf

Alonso S, Palomo A (2001a) Calorimetric study of alkaline activation of calcium hydroxide-metakaolin solid mixtures. Cem Concr Res 31:25–30. doi:10.1016/S0008-8846(00)00435-X

Alonso S, Palomo A (2001b) Alkaline activation of metakaolin and calcium hydroxide mixtures: influence of temperature, activator concentration and solids ratio. Mater Lett 47:55–62. doi:10.1016/S0167-577X(00)00212-3

Al-Rawas A, Hago A (2006) Evaluation of field and laboratory produced burnt clay pozzolans. Appl Clay Sci 31:29–35. doi:10.1016/j.clay.2005.07.009

Al-Rawas A, Hago AW, Corcoran TC (1998) Properties of Omani artificial pozzolana. Appl Clay Sci 13:275–292. doi:10.1016/S0169-1317(98)00029-5

Ambroise J, Murat M, Pera J (1985) Hydration reaction and hardening of calcined clays and related minerals. Extension of the research and general conclusions. Cem Concr Res 15:261–268

Ambroise J, Murat M, Pera J (1986) Investigation of synthetic binders obtained by middle temperature thermal dislocations of clay minerals. Silicate Indust 7:99–107

Andersson R, Gram H (1988) Properties of alkali-activated slag. In: Alkali-activated slag, Swedish cement and Concrete Research Institute, pp 9–63, CBI Research, Stockolm

Anwar M, Miyagawa T, Gaweesh M (2000) Using rice husk ash as a cement replacement material in concrete. Waste Manag 1:671-684. doi:10.1016/S0713-2743(00)80077-X

ASTM C-125 (2007) Standard terminology relating to concrete and concrete aggregates. ASTM International, West Conshohocken, Pennsylvania

Bakharev T (2005) Geopolymeric materials prepared using class F fly ash and elevated temperature curing. Cem Concr Res 35:1224–1232. doi:10.1016/j.cemconres.2004.06.031

Bakharev T, Sanjayan J, Cheny Y (1999) Alkali-activation of Australian slag cements. Cem Concr Res 29:113–120. doi:10.1016/S0008-8846(98)00170-7

Bakharev T, Sanjayan JG, Cheng YB (2003) Resistance of alkali-activated slag concrete to acid attack. Cem Concr Res 33:1607–1611. doi:10.1016/S0008-8846(03)00125-X

Balaha M, Badawy A, Hashish M (2007) Effect of using ground waste tire rubber as fine aggregate on the behaviour of concrete mixes. Indian J Eng Mater Sci 14:427–435

Banthia N, Sheng J (1996) Fracture toughness of micro-fiber reinforced cement composites. Cem Concr Compos 18:215–269. doi:10.1016/0958-9465(95)00030-5

Bankowski P, Zou L, Hodges R (2004) Using inorganic polymer to reduce leach rates of metals from brown coal fly ash. Miner Eng 17:159–166. doi:10.1016/j.mineng.2003.10.024

Barbhuiya S, Gbagbo J, Russell M, Basheer P (2009) Properties of fly ash concrete modified with hydrated lime and silica fume. Cem Concr Res 23:3233–3239. doi:10.1016/j.conbuildmat.2009.06.001

Barbosa V, MacKenzie KJ, Thaumaturgo C (2000) Synthesis and characterisation of materials based on inorganic polymers of alumina and sílica: sodium polysialate polymers. Inter J Inor Polym 2:309–317. doi:10.1016/S1466-6049(00)00041-6

Bargaheiser K, Nordmeyer D (2008) Greening of mortars with pozzolans. ASTM Special Tech Publ 1496:147–155

Batayneh M, Marie I, Asi I (2007) Use of selected waste materials in concrete mixes. Waste Manag 27:1870–1876. doi:10.1016/j.wasman.2006.07.026

Bentur A (2002) Cementitious materials-Nine millennia and a new century: Past, present, and future. J Mater Civil Eng 14:2–22. doi:10.1061/(ASCE)0899-1561

Berndt M (2009) Properties of sustainable concrete containing fly ash, slag and recycled concrete aggregate. Constr Build Mater 23:2606–2613. doi:10.1016/j.conbuildmat.2009.02.011

Bleischwitz R, Bahn-Walkowiak B (2007) Aggregates and construction markets in Europe: towards asectoral action plan on sustainable resource management. Miner Energ 22:159–176

Bogue RH (1955) The chemistry of Portland cement. Reinhold Publication Corp, New York

Bortnovsky O, Dvorakova K, Roubicek P, Bousek J, Prudkova Z, Baxa P (2007) Development, properties and production of geopolymers based on secondary raw materials. Alkali activated materials-research, production and utilization 3rd conference, pp 83–96, Prague, Czech Republic

Bouzoubaa N, Zhang M, Bilodeau A, Malhotra V (1997) The effect of grinding on the physical properties of fly ashes and a Portland cement clinker. Cem Concr Res 27:1861–1874. doi: 10.1016/j.powtec.2009.08.020

Branco F, Reis M, Tadeu A (2006) Using cork as aggregates for concrete. Meeting on quality and inovation in the construction industry. LNEC, Lisbon

BRGM (2001) Management of mining, quarrying and ore-processing waste in the European Union. European Commission, DG environment, 50319-FR http://ec.europa.eu/environment/waste/studies/mining/0204finalreportbrgm.pdf. Accessed 3 July 2011

Brough AR, Atkinson A (2002) Sodium silicate-based alkali-activated slag mortars—Part I. Strength, hydration and microstructure. Cem Concr Research 32:865–879. doi:10.1016/S0008-8846(02)00717-2

Buj I, Torras J, Casellas D, Rovira M, de Pablo J (2009) Effect of heavy metals and water content on the strength of magnesium phosphate cements. J Hazard Mater 170:345–350. doi:10.1016/j.jhazmat.2009.04.091

Bukowska M, Pacewska B, Wilińska I (2003) Corrosion resistance of cement mortars containing spent catalyst of fluidized bed cracking (FBCC) as an additive. J Therm Anal Calorim 74:931–942. doi:10.1023/B:JTAN.0000011025.26715.f5

Cairns R, Kew H, Kenny M (2004) The use of recycled rubber tyres in concrete construction. Final report, The Onyx Environmental Trust, University of Strathclyde, Glasgow

Cao J, Chung D (2002) Damage evolution during freeze–thaw cycling of cement mortar studied by electric resistivity measurement. Cem Concr Res 32:1657–1661. doi:10.1016/S0008-8846(02)00856-6

Castellanos N, Agredo J (2010) Using spent fluid catalytic cracking (FCC) catalyst as pozzolanic addition—A review. Ingen Inv 30:35–42

Chen B, Wu K, Yao W (2004) Conductivity of carbon fiber reinforced cement-based composites. Cem Concr Compos 26:291–297. doi:10.1016/S0958-9465(02)00138-5

Chen C, Habert G, Bouzidi Y, Julien A (2010) Environmental impact of cement production: detail of the different processes and cement plant variability evaluation. J Cleaner Prod 18:478–485. doi:10.1016/j.jclepro.2009.12.014

Cheng TW, Chiu JP (2003) Fire resistant geopolymer produced by granulated blast furnace slag. Miner Eng 16:205–210. doi:10.1016/S0892-6875(03)00008-6

Chmielewska B (2007) Adhesion strength and other mechanical properties of SBR modified concrete. Twelth international congress on polymers in concrete, pp 157–166, Chuncheon, Korea

Choi Y, Moon D, Chung J, Cho S (2005) Effects of waste PET bottles aggregate on the properties of concrete. Cem Concr Res 35:776–781. doi:10.1016/j.cemconres.2004.05.014

Choi Y, Kim Y, Choi O, Lee K, Lachemi M (2009) Utilization of tailings from tungsten mine waste as a substitution material for cement. Constr Build Mater 23:2481–2486. doi:10.1016/j.conbuildmat.2009.02.006

Chou L, Lin C, Lu C, Lee C, Lee M (2010) Improving rubber concrete by waste organic sulfur compounds. Waste Manag Res 28:29–35. doi:10.1177/0734242X09103843

Chung D (2000) Cement reinforced with short carbon fibers: a multifunctional material. Compos B Eng 31:511–526. doi:10.1016/S1359-8368(99)00071-2

Chung D (2001) Electromagnetic shielding effectiveness of carbon materials. Carbon 39:279–285. doi:10.4028/www.scientific.net/AMR.168-170.1438

Chung D (2002) Electric conduction behavior of cement–matrix composites. J Mater Eng Perform 11:194–204. doi:10.1361/105994902770344268

Coatanlem P, Jauberthie R, Rendell F (2006) Lightweight wood chipping concrete durability. Constr Build Mater 20:776–781. doi:10.1016/j.conbuildmat.2005.01.057

Commission European (1999) Council Directive 1999/31/EC of 26 April 1999 on the landfill of waste. Off J Eur Communities L182:1–19

Commission European (2000) Directive 2000/76/EC of the European Parliament and of the council of 4 December 2000 on incineration of waste. Off J Eur Communities L332:91–111

Corinaldesi V, Moriconi G (2009) Influence of mineral additions on the performance of 100% recycled aggregate concrete. Constr Build Mater 23:2869–2876. doi:10.1016/j.conbuildmat.2009.02.004

Coutinho A (1988) Production and properties of concrete. National laboratory of Civil Engineering, Lisbon

Criado M, Palomo A, Fernandez-Jimenez A (2005) Alkali-activation of fly ashes. Part 1: effect of curing conditions on the carbonation of the reaction products. Fuel 84:2048–2054. doi:10.1016/j.fuel.2005.03.030

Cuberos A, De la Torre A, Martín-Sedeño M, Moreno-Real L, Merlini M, Ordónez L, Aranda M (2009) Phase development in conventional and active belite cement pastes by Rietveld analysis and chemical constraints. Cem Concr Res 39:833–842. doi:10.1016/j.cemconres.2009.06.017

Dalinaidu A, Das B, Singh D (2007) Methodology for rapid determination of pozzolanic activity of materials. J ASTM Int 4. doi:10.1520/JAI100343

Damtoft J, Lukasik J, Herfort D, Sorrentino D, Gartner E (2008) Sustainable development and climate change initiatives. Cem Concr Res 38:115–127. doi:10.1016/j.cemconres.2007.09.008

Das S, Yudhbir (2006) A simplified model for prediction of pozzolanic characteristics of fly ash, based chemical composition. Cem Concr Res 36:1827–1832. doi:10.1016/j.cemconres.2006.02.020

Davidovits J (1979) Synthesis of new high temperature geo-polymers for reinforced plastics/composites. SPE PACTEC 79 Society of Plastic Engineers, pp 151–154, Brookfield Center

Davidovits J (1994a) Geopolymers: man-made rock geosynthesis and the resulting development of very early high strength cement. J Mater Educ 16:91–139

Davidovits J (1994b) Properties of geopolymers cements. In: Proceedings of the 1st International conference on alkaline cements and concretes. Scientific Research Institute on Binders and Materials Kiev, Ukraine, pp 131–149

Davidovits J (1999) Chemistry of geopolymeric systems terminology. In: Davidovits J, Davidovits R, James C (Eds) Proceedings of Géopolymère 99 2nd International Conference on geopolymers. Geopolymer Institute, France pp 9–40

Davidovits J, Sawyer JL (1985) Early high strength mineral polymer. U.S. Patent 4.509.958

Davidovits J, Comrie DC, Paterson JH, Ritcey DJ (1990) Geopolymeric concretes for environmental protection. ACI Concr Inter 12:30–40

Davis R (1996) The impact of EU and UK environmental pressures on the future of sludge treatment and disposal. J CIWEM 10:65–69. 10.1111/j.1747-6593.1996.tb00010.x

Day R, Shi Caijun (1994) Influence of the fineness of pozzolan on the strength of lime natural-pozzolan cement pastes. Cem Concr Res 8:1485–1491. doi:10.1016/0008-8846(94)90162-7

Day K, Holtze K, Metcalfe J, Bishop C, Dutka B (1993) Toxicity of leachate from automobile tyres to aquatic biota. Chemosphere 27:665–675. doi:10.1016/0045-6535(93)90100-J

Degryse P, Elsen J, Waelkens M (2002) Study of ancient mortars from Sagalassos (Turkey) in view of their conservation. Cem Concr Res 21:1457–1463. doi:10.1016/S0008-8846(02)00807-4

Dolezal J, Skvara F, Svoboda P, Sulc R, Kopecky L, Pavlasova S, Myskova L, Lucuk M, Dvoracek K (2007) Concrete based on fly ash geopolymers. Alkali activated materials-research, production and utilization 3rd conference, Prague, Czech Republic, pp 185–197

Donatello S, Tyrer M, Cheeseman C (2010a) Comparison of test methods to assess pozzolanic activity. Cem Concr Compos 32:121–127. doi:10.1016/j.cemconcomp.2009.10.008

Donatello S, Freeman-Pask A, Tyrer M, Cheeseman C (2010b) Effect of milling and acid washing on the pozzolanic activity of incinerator sewage sludge ash. Cem Concr Compos 32:54–61. doi:10.1016/j.cemconcomp.2009.09.002

Duxson P, Van Deventer J (2009) Commercialization of geopolymers for construction—opportunities and obstacles. In: Provis J, Deventer Van (eds) Structure, processing, properties and applications. Woodhead Publishing Limited Abingtone Hall, Cambridge, UK, pp 379–400

Duxson P, Provis J, Luckey G, Van Deventer J (2007) The role of inorganic polymer technology in the development of "Green Concrete". Cem Concr Res 37:1590–1597. doi: 10.1016/j.cemconres.2007.08.018

Dyer T, Dhir R (2010) Evaluation of powdered glass cullet as a means of controlling harmful alkali-silica reaction. Mag Concr Res 62:749–759

EDG (2004) European Union waste policy, LIFE Focus: a cleaner, greener Europe 3–6. ISBN 92-894-6018-0, ISSN 1725-5619. Luxembourg: Office for Official Publications of the European Communities

Eleazer W, Barlaz M, Whittle D (1992) Resource recovery alternatives for waste tires in North Carolina. School of Engineering, Civil Engineering Department, NCSU, US

Elinwa A, Ejeh S (2004) Effects of the incorporation of sawdust waste incineration fly ash in cements pastes and mortars. J Asian Arch Build Eng 3:1–7. doi:10.3130/jaabe.3.1

Elinwa A, Mahmood Y (2002) Ash from timber waste as cement replacement material. Cem Concr Compos 24:219–222. doi:10.1016/S0958-9465(01)00039-7

Escalante-Garcia JI, Gorokhovsky AV, Mendonza G, Fuentes AF (2003) Effect of geothermal waste on strength and microstructure of alkali-activated slag cement mortars. Cem Concr Res 33:1567–1574. doi:10.1016/S0008-8846(03)00133-9

Escalante-Garcia JI, Mendez-Nodell J, Gorokhovsky AV, Fraire-Luna PE, Mancha-Molinar H, Mendoza-Suarez G (2002) Reactivity and mechanical properties of alkali activated blast furnace slag. Bol Soc Esp Ceram Vidrio 41:451–458

Etxeberria M, Mari A, Vazquez E (2007) Recycled aggregate concrete as structural material. Mater Struct 40:529–541. doi:10.1617/s11527-006-9161-5

Evangelista L, Brito J (2007) Mechanical behavior of concrete made with fine recycled concrete aggregates. Cem Concr Compos 29:397–401. doi:10.1016/j.cemconcomp.2006.12.004

Fajardo P, Valdez J (2009) Corrosion of steel rebar embedded in natural pozzolan based mortars exposed to chlorides. Constr Build Mater 23:768–774. doi:10.1016/j.conbuildmat.2008.02.023

Federico L, Chidiac S (2009) Waste glass as a supplementary cementitious material in concrete-Critical review of treatment methods. Cem Concr Compos 31:606–610. doi:10.1016/j.cemconcomp.2009.02.001

Feng Q, Yamamichi H, Shova M, Sugita S (2004) Study on the pozzolanic properties of rice husk ash by hydrochloric acid pretreatment. Cem Concr Res 34:521–526. doi:10.1016/j.cemconres.2003.09.005

Fernandez R, Martirena F, Scrivener K (2011) The origin of the pozzolanic activity of calcined clay minerals: a comparison between kaolinite, illite and montmorillonite. Cem Concr Res 41:113–122. doi:10.1016/j.cemconres.2010.09.013

Fernandez-Jimenez A, Palomo A (2003) Characterization of fly ashes. Potential reactivity as alkaline cements. Fuel 82:2259–2265. doi:10.1016/S0016-2361(03)00194-7

Fernandez-Jimenez A, Palomo A (2005) Composition and microstructure of alkali activated fly ash binder: effect of the activator. Cem Concr Res 35:1984–1992

Fernandez-Jimenez A, Puertas F (1997) Alkali activated slag cements: kinetic studies. Cem Concr Res 27:359–368. doi:10.1016/S0008-8846(97)00040-9

Fernandez-Jimenez A, Palomo J, Puertas F (1999) Alkali activated slag mortars. Mechanical strength behaviour. Cem Concr Res 29:1313–1321. doi:10.1016/S0008-8846(99)00154-4

Ferreira R, Jalali S (2006) Quality control based on electrical resistivity measurements. In: Proceedings of the European symposium on service life and serviceability of the concrete structures, Helsink, Finland

Flower D, Sanjayan J (2007) Green house gas emissions due to concrete manufacture. Inter J Life Cycle Assess 12:282–288

Foladori P, Gianni A, Ziglio G (2010) Sludge reduction technologies in wastewater treatment plants. IWA Publishing, London

Fowler DW (1999) Polymers in concrete: a vision for the 21st century. Cem Concr Compos 21:449–452

Franke L, Sisomphon K (2004) A new chemical method for analyzing free calcium hydroxide content in cementing material. Cem Concr Res 34:1161–1165. doi:10.1016/j.cemconres.2003.12.003

Frias M, Villar-Cocina E, Sanchez de Rojas M, Valencia-Morales E (2005) The effect that different pozzolanic activity methods has on the kinetic constants of the pozzolanic reaction in sugar cane straw-clay ash/lime systems: application of a kinetic–diffusive model. Cem Concr Res 35:2137–2142. doi:10.1016/j.cemconres.2005.07.005

Fytili D, Zabaniotou A (2008) Utilization of sewage sludge in EU application of old and new methods—A review. Renew Sustainable Energy Rev 12:116–140. doi:10.1016/j.cemconcomp.2009.02.001

Ganjian E, Khorami M, Maghsoudi A (2009) Scrap-tyre-rubber replacement for aggregate and filler in concrete. Constr Build Mater 23:1828–1836

Gartner E (2004) Industrially interesting approaches to low-CO_2 cements. Cem Concr Res 34:1489–1498

Gava G, Prudêncio L (2007a) Pozzolanic activity tests as a measure of pozzolans performance. Part 1. Mag Concr Res 59:729–734

Gava G, Prudêncio L (2007b) Pozzolanic activity tests as a measure of pozzolans performance. Part 2. Mag Concr Res 59:735–741

Ghaly A, Cahill J (2005) Correlation of strength, rubber content, and water to cement ratio in rubberized concrete. Can J Civil Eng 32:1075–1081

Gjorv OE (1992) High strength concrete. In: Malhotra VM (ed) Advances in concrete technology, American Concrete Institute Montreal Canada, pp 21–77

Glavind M (2009) Sustainability of cement, concrete and cement replacement materials in construction. In: Khatib J (ed) Sustainability of Construction Materials. WoodHead Publishing in Materials, Great Abington, Cambridge, pp 120–147

Glukhovsky V (1959) Soil silicates. Gostroiizdat Publish, Kiev, USSR

Glukhovsky VD, Rostovskaja GS, Rumyna GV (1980) High strength slag alkaline cements. 7th International congress on the chemistry of cement

Gnecco L (1999) Building a shield room is not construction. Evaluation Eng 38

Gore A (2009) Our choice. A plan to solve the climatic crisis. Rodale Books, Emaus

Gourley JT (2003) Geopolymers:opportunities for environmentally friendly construction materials. Materials 2003 Conference, Institute of Materials Engineering Australasia

Gourley JT, Johnson GB (2005) Developments in geopolymer precast concrete. In: Proceedings of Geopolymer 2005 World congress, geopolymer green chemistry and sustainable development solutions, S. Quentin, France, pp 139–143

Granizo ML (1998) Activation alcalina de metacaolin: Desarrolllo de nuevos materials cementantes. PhD thesis University Autoneoma of Madrid

Guleç A, Tulun A (1997) Physico-chemical and petrographical studies of old mortars and plasters of Anatolia. Cem Concr Res 27:227–234. doi:10.1016/S0008-8846(97)00005-7

Guneyisi E, Gesoglu M, Ozturan T (2004) Properties of rubberized concretes containing silica fume. J Cem Concr Res 34:2309–2317. doi:10.1016/j.cemconres.2004.04.005

Hannawi K, Kamali-Bernard S, Prince W (2010) Physical and mechanical properties of mortars containing PET and PC waste aggregates. Waste Manag 30:2312–2320. doi:10.1016/j.wasman.2010.03.028

Hardjito D, Wallah SE, Sumajouw D, Rangan BV (2002a) Research into engineering properties of geopolymer concrete. In: Proceedings of 2002 Geopolymer conference. Melbourne, Australia

Hardjito D, Wallah SE, Sumajouw D, Rangan BV (2002b) Properties of geopolymer concrete with fly ash source material: effect of mixture composition. In: Proceedings of seventh CANMET/ACI International conference on recent advances in concrete technology, Las Vegas, USA

Hazra PC, Krishnaswamy VS (1987) Natural pozzolans in India, their utility, distribution and petrogragraphy. Rec Geol Surv India 87:675–706

He C, Makovic E, Osbaeck B (1995a) Thermal stability and pozzolanic activity of raw and calcined illite. Appl Clay Sci 9:337–354.doi:10.1016/0169-1317(94)00033-M

He C, Osbaeck B, Makovicky E (1995b) Pozzolanic reactions of six principal clay minerals: activation, reactivity assessments and technological effects. Cem Concr Res 25:1691–1702.doi:10.1016/0008-8846(95)00165-4

He C, Makovic E, Osbaeck B (2000) Thermal stability and pozzolanic activity of raw and calcined mixed-layer mica/smectite. Appl Clay Sci 17:141–161.doi:10.1016/S0169-1317(00)00011-9

Hermann E, Kunze C, Gatzweiler R, Kiebig G, Davidovits J (1999) Solidification of various radioactive residues by geopolymere with special emphasis on long term stability. In: Proceedings of 1999 geopolymere conference, pp 211–228

Hossain M, Lachemi M, Sahmaran M (2009) Performance of cementitious building renders incorporating natural and industrial pozzolans under aggressive airborne marine salts. Cem Concr Compos 31:358–368.doi:10.1016/j.cemconcomp.2009.03.005

Houget V (1992) Etude dês caracteristiques mecaniques et physico-chimiques de composites ciments-fibres organiques. Ph.D. dissertation, Inst Nat Sci Appl, Lyon, France

Hu S, Wang H, Zhang G, Ding Q (2008) Bonding and abrasion resistance of geopolymeric repair material made with steel slag. Cem Concr Compos 30:239–244. doi:10.1016/j.cemconcomp.2007.04.004

Huntzinger D, Eatmon T (2009) A life-cycle assessment of Portland cement manufacturing: comparing the traditional process with alternative technologies. J Cleaner Prod 17:668–675. doi:10.1016/j.jclepro.2008.04.007

Jiang L, Liu Z, Ye Y (2004) Durability of concrete incorporating large volumes of low-quality fly ash. Cem Concr Res 34:1467–1469. doi:10.1016/j.cemconres.2003.12.029

Josa A, Aguado A, Cardim A, Byars E (2007) Comparative analysis of the life cycle impact assessment of available cement inventories in the EU. Cem Concr Res 37:781–788. doi: 10.1016/j.cemconres.2007.02.004

Kaid N, Cyr M, Julien S, Khelafi H (2009) Durability of concrete containing a natural pozzolan as defined by a performance-based approach. Constr Build Mater 23:3457–3467. doi:10.1016/j.conbuildmat.2009.08.002

Katz A (1998) Microscopic study of alkali-activation fly ash. Cem Concr Res 28:197–208

Khatib J (2009) Sustainability of construction materials. WoodHead Publishing in Materials, Cambridge

Kiattikomol K, Jaturapitakkul C, Songpiriyakij S, Hutubtim S (2001) A study of ground coarse fly ashes with different fineness from various sources as pozzolanic materials. Cem Concr Compos 21:335–343. doi:10.1016/S0958-9465(01)00016-6

Kim J, Park C, Lee S, Lee S, Won J (2008) Effects of the geometry of recycled PET fibre reinforcement on shrinkage cracking of cement-based composites. Compos B 39:442–450. doi:10.1016/j.compositesb.2007.05.001

Kim S, Yi N, Kim H, Kim J, Song Y(2010) Material and structural performance evaluation of recycled PET fiber reinforced concrete. Cem Concr Compos 32:232–240. doi: 10.1016/j.cemconcomp.2009.11.002

Kirschner A, Harmuth H (2002) Investigation of geopolymer binders with respect to their application for building materials. Ceramics–Silicaty 48:117–120. http://www.ceramics-silikaty.cz/2004/pdf/2004_03_117.pdf

Kong D, Sanjayan J, Sagoe-Cretensil K (2008) Factors affecting the performance of metakaolin geopolymers exposed to elevated temperatures. J Mater Sci 43:824–831. 10.1007/s10853-007-2205-6

Krivenko P, Guziy S (2007) Fire resistant alkaline portland cements. Alkali activated materials—research, production and utilization 3rd conference, Prague, Czech Republic, pp 333–347

Krizan D, Zivanovic B (2002) Effects of dosage and modulus of water glass on early hydration of alkali-slag cements. Cem Concr Res 32:1181–1188. doi:10.1016/S0008-8846(01)00717-7

Lachemi M, Jagnit-Hamou A, Aïtcin C (1998) Long-term performance of silica fume cement concretes. Concr Inst 20:59–65

Lancellotti I, Kamseu E, Michelazzi M, Barbieri L, Corradi A, Leonelli C (2010) Chemical stability of geopolymers containing municipal solid waste incinerator fly ash. Waste Manag 30:673–679. doi:10.1016/j.wasman.2009.09.032

Laukaitis A, Zurauskas R, Keriené J (2005) The effect of foam polystyrene granules on cement composites properties. Cem Concr Compos 27:41–47. doi:10.1016/j.cemconcomp.2003.09.004

Lavat A, Trezza M, Poggi M (2009) Characterization of ceramic roof tile wastes as pozzolanic admixture. Waste Manag 29:1666–1674. doi:10.1016/j.wasman.2008.10.019

Lea FM (1970) The chemistry of cement, 3rd edn. Edward Arnold ltd., London
Lee W, Van Deventer J (2002a) The effect of ionic contaminants on the early-age properties of alcali-activated fly ash-based cements. Cem Concr Res 32:577–584. doi:10.1016/S0008-8846(01)00724-4
Lee W, Van Deventer J (2002b) The effects of inorganic salt contamination on the strength and durability of geopolymers. Coll Surf 211:115–126. doi:10.1016/S0927-7757(02)00239-X
Li G, Garrick G, Eggers J, Abadie C, Stubblefield M, Pang S (2004) Waste tire fiber modified concrete. Compos B35:305–312. doi:10.1016/j.compositesb.2004.01.002
Lundin M, Olofsson M, Pettersson G, Zetterlund H (2004) Environmental and economic assessment of sewage sludge handling options. Res Cons Rec 41:255–278. doi:10.1016/j.resconrec.2003.10.006
Luxán M (1976) Estudio de las puzolanas de origen volcánico mediante espectroscopia de absorción infrarroja-Cuadernos de Investigación. Instituto EduardoTorroja 32:5–21
Luxan M, Madruga F, Saavedra J (1989) Rapid evaluation of pozzolanic activity of natural products by conductivity measurement. Cem Concr Res 19:63–68. doi:10.1016/0008-8846(89)90066-5
Malerious O (2003) Modelling the adsorption of mercury in the flue gas of sewage sludge incineration. Chem Eng J 96:197–205. doi:10.1016/j.cej.2003.08.018
Malhotra V(1978) Recycled concrete: a new aggregate. Can J Civil Eng 5:42–52
Malhotra V (2007) Global warning and sustainability issues related to concrete technology. In: Proceedings of the International conference on sustainability in the cement and concrete industry, Lillehammer, Norway
Malhotra V, Mehta P (1996) Pozzolanic and cementitious materials. Gordon and Breacch Publisher, Canada
Malhotra V, Mehta P (2005) High performance, high-volume fly ash concrete: materials, mixture proportioning, properties, construction practice, and case histories. Supplementary Cementing Materials for Sustainable Development Inc., Ottawa
Malinowsky R (1991) Prehistory of concrete. Concr Inter 13:62–68
Martín-Sedeño M, Cuberos A, De la Torre A, Álvarez-Pinazo G, Ordónez L, Gateshki M, Aranda M (2010) Aluminum-rich belite sulfoaluminate cements: clinkering and early age hydration. Cem Concr Res 40:359–369. doi:10.1016/j.cemconres.2009.11.003
Marzouk O, Dheilly R, Queneudec M (2007) Valorization of post-consumer waste plastic in cementitious concrete composites. Waste Manag 27:310–318. doi:10.1016/j.wasman.2006.03.012
McCarter W, Tran D (1996) Monitoring pozzolanic activity by direct activation with calcium hydroxide. Constr Build Mater 10:179–184. doi:10.1016/0950-0618(95)00089-5
Mccartthy M, Dhir R (1999) Towards maximising the use of fly ash as a binder. Fuel 78:121–123. doi:10.1016/S0016-2361(98)00151-3
Mccartthy M, Dhir R (2004) Development of a high volume fly ash cements for use in concrete construction. Fuel 84:1423–1432. doi:10.1016/j.fuel.2004.08.029
Mehta K (1998) Role of pozzolanic and cementitious material in sustainable development of the concrete industry. In: Proceedings of the 6th International Conference on the use of fly ash, silica fume, slag and natural pozzolans in concrete. SP 178 ACI International, pp 1–25
Mehta K (2001) Reducing the environmental impact of concrete. Concr Intern 61–66. http://www.ecosmartconcrete.com/kbase/filedocs/trmehta01.pdf
Mello D, Pezzin S, Amico S (2009) The effect of post consumer PET particles on the performance of flexible polyurethane foams. Polym Test 28:702–708. doi:10.1016/j.polymertesting.2009.05.014
Menezes R, Ferreira H, Neves A, Ferreira H (2003) Characterization of ball clays from the coastal region of Paraiba state. Ceramica 49:120–127. doi:10.1590/S0366-69132003000300003
Meyer C (2009) The greening of the concrete industry. Cem Concr Compos 31:601–605. doi:10.1016/j.cemconcomp.2008.12.010

Monteny J, Belie N, Vincke E, Verstraete W, Taerwe L (2001) Chemical and microbiological tests to simulate sulfuric acid corrosion of polymer-modified concrete. Cem Concr Res 31:1359–1365. doi:10.1016/S0008-8846(01)00565-8

Monzo J, Paya J, Borrachero MV, Peris-Mota E (1999) Mechanical behavior of mortars containing sewage sludge ash (SSA) and Portland cements with different tricalcium aluminate content. Cem Concr Res 29:87–94. doi:10.1016/S0008-8846(98)00177-X

Mora E (2007) Life cycle, sustainability and the transcendent quality of building materials. Build Environ 42:1329–1334. doi:10.1016/j.buildenv.2005.11.004

Moropoulou A, Bakolas A, Aggelakopoulou E (2004) Evaluation of pozzolanic activity of natural and artificial pozzolans by thermal analysis. Thermochim Acta 420:135–140. doi: 10.1016/j.tca.2003.11.059

Morsli K, De la Torre A, Zahir M, Aranda M (2007) Mineralogical phase analysis of alkali and sulfate bearing belite rich laboratory clinkers. Cem Concr Res 37:639–646. doi:10.1016/j.cemconres.2007.01.012

Müller I (2004) Influence of silica fume addition on concretes physical properties and on corrosion behavior of reinfor cement bars. Cem Concr Comp 26:31–39. doi: 10.1016/S0958-9465(02)00120-8

Mullick A, Babu K, Rao P (1986) Evaluation of pozzolanic activity and its impact on specification of blended cements. In: Proceedings of 8th international congress on the chemistry of cement, vol VI, pp 308–311

Naceri A, Hamina M (2009) Use of waste brick as a partial replacement of cement in mortar. Waste Manag 29:2378–2384. doi:10.1016/j.wasman.2009.03.026

Nagdi K (1993) Rubber as an engineering material: guidelines for user. Hanser Publication, Cincinnati

Naik T, Singh S (1991) Utilization of discarded tyres as construction materials for transportation facilities. Report N CBU-1991-02, UWM Center for by-products utilization. University of Wiscosin, Milwaukee

Naik T, Singh S, Wendorf R (1995) Applications of scrap tire rubber in asphaltic materials: state of the art assessment. Report N CBU-1995-02, UWM Center for by-products utilization. University of Wiscosin, Milwaukee

Neelamegan M, Dattatreya J, Harish K (2007) Effect of latex and fibber addition on mechanical and durability. Properties of sintered fly ash lightweight aggregate concrete mixtures. In: Proceedings of 12th International Congress on Polymers in Concrete, Chuncheon, Korea, pp 113–121

Neville AM (1997) Properties of concrete, 4th edn. Wiley, New York

Ochi T, Okubo S, Fukui K (2007) Development of recycled PET fibre and its application as concrete-reinforcing fibre. Cem Concr Compos 29:448–455. doi:10.1016/j.cemconcomp.2007.02.002

Ogawa H, Kano K, Mimura T, Nagai K, Shirai A, Ohama Y (2007) Durability performance of barrier penetrants on concrete surfaces. In: Proceedings of 12th International congress on polymers in concrete, Chuncheon, Korea, pp 373–382

Ohama Y (1998) Polymer-based admixtures. Cem Concr Compos 20:189–212. doi: 10.1016/S0958-9465(97)00065-6

Ohama Y (2010) Concrete–polymer composites: the past, present and future. 12th International Congress on Polymers on Concrete, Madeira, Portugal, pp 1–13

Oiknomou N, Stefanidou M, Mavridou S (2006) Improvement of the bonding between rubber tire particles and cement paste in cement products. 15th Conference of the technical chamber of greece, Alexandroupoli, Greece, pp 234–242

Oikonomou N, Mavridou S (2009) The use of waste tyre rubber in civil engineering works. In: Khatib J (ed) Sustainability of construction materials. WoodHead Publishing Limited, Abington Hall, Cambridge

Oliveira L, Jalali S, Fernandes J, Torres E (2005) L′emploi de métakaolin dans la production de béton écologiquement efficace. Mater Struct 38:403–410. doi:10.1617/14186

Ozkan S, Gjorv O (2008) Electrical resistivity measurements for the quality control during concrete construction. ACI Mater J 105:541–547

Pacewska B, Bukowska M, Wilińska I, Swat M (2002) Modification of the properties of concrete by a new pozzolan—A waste catalyst from the catalytic process in a fluidized bed. Cem Concr Res 32:145–152. doi:10.1016/S0008-8846(01)00646-9

Pacheco-Torgal F (2007) Development of alkali-activated binders based on mine waste mud form the Panasqueira mine. Ph.D. thesis, UBI, Covilhã, Portugal

Pacheco-Torgal F, Jalali S (2009) sulfuric acid resistance of plain, polymer modified, and fly ash cement concretes. Constr Build Mater 23:3485–3491. doi:10.1016/j.conbuildmat.2009.08.001

Pacheco-Torgal F, Jalali S (2010) Reusing ceramic wastes in concrete. Constr Build Mater 24:832–838. doi:10.1016/j.conbuildmat.2009.10.023

Pacheco-Torgal F, Jalali S (2011) Using metakaolin to improve the sustainability of fly ash based concrete. Intern J Sustainable Eng

Pacheco-Torgal F, Gomes JP, Jalali S (2005) Geopolymeric binder using tungsten mine waste: preliminary investigation. In: Proceedings of Geopolymer 2005 World congress, S. Quentin, France, pp 93–98

Pacheco-Torgal F, Gomes JP, Jalali S (2006) Bond strength between concrete substrate and repair materials. Comparisons between Tungsten mine waste geopolymeric binder versus current commercial repair products. In: Proceedings of the 7th International congress on advances in civil engineering, Turquey, p 482

Pacheco-Torgal F, Gomes JP, Jalali S (2007a) Durability and environmental performance of alkali-activated tungsten mine waste mud mortars. J Mater Civil Eng 22:897–904. 10.1061/(ASCE)MT.1943-5533.0000092

Pacheco-Torgal F, Gomes JP, Jalali S (2007b) Investigations about the effect of aggregates on strength and microstructure of geopolymeric mine waste mud binders. Cem Concr Res 37:933–941. doi:10.1016/j.cemconres.2007.02.006

Pacheco-Torgal F, Gomes JP, Jalali S (2008a) Investigations on mix design of tungsten mine waste geopolymeric binders. Constr Build Mater 22:1939–1949. doi:10.1016/j.conbuildmat. 2007.07.015

Pacheco-Torgal F, Gomes J P, Jalali S (2008b) Properties of tungsten mine waste geopolymeric binder. Constr Build Mater 22:1201–1211. doi:10.1016/j.conbuildmat.2007.01.022

Pacheco-Torgal F, Gomes JP, Jalali S (2008c) Adhesion characterization of tungsten mine waste geopolymeric binder. Influence of OPC concrete substrate surface treatment. Constr Build Mater 22:154–161. doi:10.1016/j.conbuildmat.2006.10.005

Pacheco-Torgal F, Gomes JP, Jalali S (2009a) Tungsten mine waste geopolymeric binders. Preliminary hydration products. Constr Build Mater 23:200–209. doi:10.1016/j.conbuildmat. 2008.01.003

Pacheco-Torgal F, Gomes JP, Jalali S (2009b) Utilization of mining wastes to produce geopolymer binders. In: Provis J, Van Deventer J (eds) Geopolymers, structure, processing, properties and applications. Woodhead Publishing Limited, Abington Hall, Cambridge

Palmer W (2010) The fly ash threat. Concr Prod 28:29–34

Palomo A, Palacios M (2003) Alkali-activated cementitious materials: alternative matrices for the immobilisation of hazardous wastes Part II. Stabilisation of chromium and lead. Cem Concr Res 33:289–295. doi:10.1016/S0008-8846(02)00964-X

Palomo A, Blanco-Varela MT, Granizo ML, Puertas F, Vasquez T, Grutzeck MW (1999a) Chemical stability of cementitious materials based on metakaolin. Cem Concr Res 29:997–1004. doi:10.1016/S0008-8846(99)00074-5

Palomo A, Grutzek MW, Blanco MT (1999b) Alkali-activated fly ashes. A cement for the future. Cem Concr Res 29:1323–1329. doi:10.1016/S0008-8846(98)00243-9

Pan S, Tseng D, Lee C (2002) Use of sewage sludge ash as fine aggregate and pozzolan in portland cement mortar. J Solid Waste Technol Manag 28:121–130

Panyakapo P, Panyakapo M (2008) Reuse of thermosetting plastic waste for lightweight concrete. Waste Manag 28:1581–1588. doi:10.1016/j.wasman.2007.08.006

Papadakis V, Fardis M, Vayenas C (1992) Hydration and carbonation of pozzolanic cements. ACI Mater J 89(2):119–130

Park S, Lee B (1993) Mechanical properties of carbon fiber-reinforced polymer-impregnated cement composites. Cem Concr Compos 15:153–163. doi:10.1016/0958-9465(93)90004-S

Pawlasova S, Skavara F (2007) High-temperature properties of geopolymer materials. Alkali activated materials-research, production and utilization 3rd conference, Prague, Czech Republic, pp 523–524

Paya J, Borrachero M, Monzo J, Peris-Mora E, Amahjour F (2001) Enhanced conductivity measurements techniques for evaluation of fly ash pozzolanic activity. Cem Concr Res 31:41–49. doi:10.1016/S0008-8846(00)00434-8

Payá J, Borrachero M, Monzó J, Soriano L (2009) Studies on the behavior of different spent fluidized-bed catalytic cracking catalysts on Portland cement. Mater Constr 59:37–52

Perná I, Hanzlicek T, Straka P, Steinerova M (2007) Utilization of fluidized bed ashes in thermal resistance applications. Alkali activated materials-research, production and utilization 3rd conference, Prague, Czech Republic, pp 527–537

Phair J, Smith J, Van Deventer J (2004) Effect of Al source and alkali-activation on Pb and Cu immobilisation in fly ash-based geoplymers. Appl Geochem 19:423–434

Pigeon M, Azzabi M, Pleau R (1996a) Can microfibers prevent frost damage? Cem Concr Res 26:1163–1170. doi:10.1016/0008-8846(96)00098-1

Pigeon M, Pleau R, Azzabi M, Banthia N (1996b) Durability of microfiber-reinforced mortars. Cem Concr Res 26:1163–1170. doi:10.1016/0008-8846(96)00015-4

Pinto AT (2004) Alkali-activated metakaolin based binders. Ph.D. thesis, University of Minho, Portugal

Popescu C, Muntean M, Sharp J (2003) Industrial trial production of low energy belite cement. Cem Concr Compos 25:689–693. doi:10.1016/S0958-9465(02)00097-5

Pourkhorshidi A, Najimi M, Parhizkar T, Hillemeier B, Herr R (2010a) A comparative study of the evaluation methods for pozzolans. Adv Cem Res 22:157–164. doi:10.1680/adcr.2010.22.3.157

Pourkhorshidi A, Najimi M, Parhizkar T, Jafarpour F, Hillemeier B (2010b) Applicability of the standard specifications of ASTM C618 for evaluation of natural pozzolans. Cem Concr Compos 32:794–800. doi:10.1016/j.cemconcomp.2010.08.007

Puertas F, Fernandez-Jimenez A (2003) Mineralogical and microstrutural characterisation of alcali-activated fly ash/slag pastes. Cem Concr Compos 25:287–292. doi:10.1016/S0958-9465(02)00059-8

Puertas F, Martinez-Ramirez S, Alonso S, Vasquez T (2000) Alkali-activated fly ash/slag cement. Strength behaviour and hydration products. Cem Concr Res 30:1625–1632. doi:10.1016/S0008-8846(00)00298-2

Puertas F, Garcia-Diaz I, Barba A, Gazulla M, Palacios M, Gomez M, Martinez-Ramirez S (2008) Ceramic wastes as alternative raw materials for Portland cement clinker production. Cem Concr Compos 30:798–805. doi:10.1016/j.cemconcomp.2008.06.003

Purdon AO (1940) The action of alkalis on blast furnace slag. J Soc Chem Ind 59:191–202

Qian G, Sun D, Tay J (2003) Characterization of mercury and zinc-doped alkali-activated slag matrix. Part II. Zinc. Cem Concr Res 33:1271–1262. doi:10.1016/S0008-8846(03)00046-2

Qiao F, Chau C, Li Z (2009) Setting and strength development of magnesium phosphate cement paste. Adv Mater Res 21:175–180. 10.1680/adcr.9.00003

Qiao F, Chau C, Li Z (2010) Property evaluation of magnesium phosphate cement mortar as patch repair material. Constr Build Mater 24:695–700. doi:10.1016/j.conbuildmat.2009.10.039

Quillin K (2001) Performance of belite–sulfoaluminate cements. Cem Concr Res 31:1341–1349. doi:10.1016/S0008-8846(01)00543-9

Raghavan D, Huynh H, Ferraris C (1998) Workability, mechanical properties, and chemical stability of a recycled tyre rubber filled cementitious composite. J Mater Sci 33:1745–1752. doi:10.1023/A:1004372414475

Rahhal V (2002) Characterization of pozzolanic additions by conduction calorimetry. Ph.D. Thesis, Politechnic University of Madrid, E.T.S. Ings. Caminos, Canales y Puertos
Rahhal V, Talero R (2010) Fast physics-chemical and calorimetric characterization of natural pozzolans and other aspects. J Therm Anal Calorim 99:479–486. doi:10.1007/s10973-009-0016-5
Rahier H, Van Melle B, Biesemans M, Wastiels J, Wu X (1996) Low temperature synthesized aluminosilicate glasses Part I. Low temperature reaction stoichimetry and structure of a model compound. J Mater Sci 31:71–79. doi:10.1007/BF00355129
Rahier H, Simons W, Van Melle B, Biesemans M (1997) Low temperature synthesized aluminosilicate glasses Part III. Influence of composition of the silica solution on production, structure and properties. J Mater Sci 32:2237–2247. doi:10.1023/A:1018563914630
Ramachandran V (1932) Concrete admixtures handbook. Properties science and technology. Noyes Publications, Park Ridge
Rassk E, Bhaskar M (1975) Pozzolanic activity of pulverized fuel ash. Cem Concr Res 5:363–376. doi:10.1016/0008-8846(75)90091-5
Reza F, Batson J, Yamamuro J, Lee J (2003) Resistant changes during compression of carbon fiber cement composites. J Mater Civil Eng 15:476–483. doi:10.1061/(ASCE)0899-1561(2003)15:5(476)
Rodriguez-Camacho R, Uribe-Afif R (2002) Importance of using the natural pozzolans on concrete durability. Cem Concr Res 32:1851–1858. doi:10.1016/S0008-8846(01)00714-1
Rosell-Lam M, Villar-Cocina E, Frias M (2011) Study on the pozzolanic properties of a natural Cuban zeolitic rock by conductometric method: kinetic parameters. Constr Build Mater 25:644–650. doi:10.1016/j.conbuildmat.2010.07.027
Roskovic R, Bjegovic D (2005) Role of mineral additions in reducing CO_2 emission. Cem Concr Res 35:974–978. doi:10.1016/j.cemconres.2004.04.028
Rossignolo J (2005) Porosity and calcium hydroxide content of portland cement pastes with active sílica and SBR látex. R Mater 10:437–442
Rowles M, O′Connor B (2003) Chemical optimisation of the compressive strength of aluminosilicate geopolymers synthetised by sodium silicate activation of metakaolinite. J Mater Chem 13:1161–1165
Roy D (1987) Hydration of blended cements containing slag, fly ash or sílica fume. In: Proceedings of Meeting Institute of Concrete Technology, Coventry, UK, pp 29–31
Roy DM, Langton C (1989) Studies of ancient concretes as analogs of cementituos sealing materials for repository in Tuff. L A-11527-MS, Los Alamos Nacional Laboratory, Los Alamos
Sabir B, Wild S, Bai J (2001) Metakaolin and calcined clays as pozzolans for concrete: a review. Cem Concr Compos 23:441–454. doi:10.1016/S0958-9465(00)00092-5
Sahayan A, Xu A (20049 Value-added utilisation of waste glass in concrete. Cem Concr Res 34:81–89. doi:10.1016/S0008-8846(03)00251-5
Sbordoni-Mora L (1981) Les matériaux des enduits traditionnels, In: Proceedings of ICCROM symposium mortars, cements and grouts used in the conservation of historic buildings, Rome, pp 375–385
Scrivener K, Kirkpatrick R (2008) Innovation in use and research on cementitious material. Cem Concr Res 38:128–136. doi:10.1016/j.cemconres.2007.09.025
Segre N, Joekes I (2000) Use of tire rubber particles as addition to cement paste. Cem Concr Res 30:1421–1425. doi:10.1016/S0008-8846(00)00373-2
Segre N, Monteiro P, Sposito G (2002) Surface characterization of recycled tire rubber to be used in cement paste matrix. J Coll Interface Sci 248:521–523. doi:10.1016/S0008-8846(00)00373-2
Shao Y, Lefort T, Moras S, Rodriguez D (2000) Studies on concrete containing ground waste glass. Cem Concr Res 30:91–100. doi:10.1016/S0008-8846(99)00213-6
Shao-Dong Wang, Scrivener K, Pratt P (1994) Factors affecting the strength of alkali-activated slag. Cem Concr Res 24:1033–1043. doi:10.1016/0008-8846(94)90026-4

Sharp J, Lawrence C, Yang R (1999) Calcium sulfoaluminate cements—low energy cements special cements or what? Adv Cem Res 11:3–13

Shi C, Day R (1995) A calorimetric study of early hydration of alkali-slag cements. Cem Concr Res 25:1333–1346. doi:10.1016/0008-8846(95)00126-W

Shirai A, Kano K, Nagai K, Ide K, Ogawa H, Ohama Y (2007) Basic properties of barrier penetrants as polymeric impregnants for concrete surfaces. 12th International congress on polymers in concrete, Chuncheon, Korea, pp 607–615

Silva D, Betioli A, Gleize P, Roman H, Gomez L, Ribeiro J (2005) Degradation of recycled PET fibers in Portland cement-based materials. Cem Concr Res 35:1741–1746. doi: 10.1016/j.cemconres.2004.10.040

Sinthaworn S, Nimityongskul P (2009) Quick monitoring of pozzolanic reactivity of waste ashes. Waste Manag 29:1526–1531. doi:10.1016/j.wasman.2008.11.010

Sisomphon K, Franke L (2011) Evaluation of calcium hydroxide contents in pozzolanic cement pastes by a chemical extraction method. Constr Build Mater 25:190–194. doi: 10.1016/j.conbuildmat.2010.06.039

Song X, Marosszeky M, Brungs M, Munn R (2005) Durability of fly ash-based geopolymer concrete against sulfuric acid attack. 10th International Conference on Durability of Building Materials and Components, Lyon, France

Song H, Pack S, Nam S, Jang J, Saraswathy V (2010) Estimation of the permeability of silica fume cement concrete. Constr Build Mater 24:315–321. doi:10.1016/j.conbuildmat. 2009.08.033

Soudée E, Péra J (2000) Mechanism of setting reaction in magnesia-phosphate cements. Cem Concr Res 30:315–321. doi:10.1016/S0008-8846(99)00254-9

Sousa Coutinho J (2003) The combined benefits of CPF and RHA in improving the durability of concrete structures. Cem Concr Comp 25:51–59. doi:10.1016/S0958-9465(01)00055-5

Sumajow M, Rangan B (2006) Low-calcium fly ash-based geopolymer concrete: reinforced beams and columns. research report GC, Curtin University of Technology, Perth, Australia

Swanepoel J, Strydom C (2002) Utilization of fly ash in a geopolymeric material. Appl Geochem 17:1143–1148. doi:10.1016/S0883-2927(02)00005-7

Talling B, Brandstetr J (1989) Present state and future of alkali-activated slag concretes. 3rd International Conference on fly ash, silica fume, slag and natural pozzolans in concrete, Trondheim, Norway, pp 1519–1546

Tashiro C, Ikeda K, Inoue Y (1994) Evaluation of pozzolanic activity by the electric resistance measurement method. Cem Concr Res 24:1333–1139. doi:10.1016/0008-8846(94)90037-X

Tiruta-Barna L, Benetto E, Perrodin Y (2007) Environmental impact and risk assessment of mineral wastes reuse strategies: Review and critical analysis of approaches and applications. Resour Conser Recy 50:351–379. doi:10.1016/j.resconrec.2007.01.009

Tyrer M, Cheeseman C, Greaves R, Claisse P, Ganjian E, Kay M, Churchman-Davies J (2010) Potential for carbon dioxide reduction from cement industry through increased use of industrial pozzolans. Adv Appl Ceram 109:275–279. doi: 10.1179/174367509X12595778633282

Uzal B, Turanli L, Yucel H, Goncuoglu M, Culfaz A (2010) Pozzolanic activity of clinoptilotite: a comparative study with silica fume, fly ash and a non-zeolitic natural pozzolan. Cem Concr Res 40:398–404. doi:10.1016/j.cemconres.2009.10.016

Van den Heede P, Gruyaerta E, De Belie N (2010) Transport properties of high-volume fly ash concrete: capillary water sorption, water sorption under vacuum and gas permeability. Cem Concr Compos 32:749–756. doi:10.1016/j.cemconcomp.2010.08.006

Van Deventer J, Provis J, Duxson P, Brice D (2010) Chemical research and climate change as drivers in the commercial adoption of alkali activated materials. Waste Biomass Valoriz 1:145–155. doi: 10.1007/s12649-010-9015-9

Van Gemert D, Knapen E (2010) Contribution of C-PC to sustainable construction procedures. International Congress on Polymers on Concrete, Madeira, Portugal, pp 27–36

Van Jaarsveld J, Van Deventer J (1999) The effect of metal contaminants on the formation and properties of waste-based geopolymers. Cem Concr Res 29:1189–1200. doi:10.1016/ S0008-8846(99)00032-0

Van Jaarsveld J, Van Deventer J, Lorenzen L (1997) The potential use of geopolymeric materials to immobilize toxic metals: Part I. Theory and applications. Min Eng 10:659–669

Van Jaarsveld J, Van Deventer J, Lorenzen L (1998) Factors affecting the immobilisation of metals in geopolymerised fly ash. Metall Mater Trans B 29:283–291

Van Jaarsveld J, Van Deventer J, Lukey GC (2003) The characterisation of source materials in fly ash-based geopolymers. Mater Lett 57(7):1272–1280. doi:10.1016/S0167-577X(02)00971-0

Vieira R, Soares R, Pinheiro S, Paiva O, Eleutério J, Vasconcelos R (2010) Completely random experimental design with mixture and process variables for optimization of rubberized concrete. Constr Build Mater 24:1754–1760. doi:10.1016/j.conbuildmat.2010.02.013

Villar-Cocina E, Valencia-Morales E, Gonzalez-Rodriguez R, Hernandez-Ruiz J (2003) Kinetics of the pozzolanic reaction between lime and sugar cane straw ash by electrical measurement: a kinetic–diffusive model. Cem Concr Res 33:517–524. doi:10.1016/S0008-8846(02)00998-5

Vinsova H, Jedinakova-Krizova G, Sussmilch J (2007) Immobilization of toxic contaminants into aluminosilicate matrixes. In: Proceedings of alkali activated materials—research, production and utilization 3rd conference, Prague, Czech Republic, pp 735–736

Wang S (1991) Review of recent research on alkali-activated concrete in China. Mag Concr Res 43:29–35. doi:10.1680/macr.1991.43.154.29

Wang Shao-Dong, Scrivener K (1995) Hydration products of alkali activated slag cement. Cem Concr Res 25:561–571. doi:10.1016/0008-8846(95)00045-E

Wang Y, Backer S, Li V (1987) An experimental study of synthetic fibre reinforced cementitious composites. J Mater Sci 22:4281–4291. doi:10.4028/www.scientific.net/AMR.150-151.1013

Wang H, Li H, Yan F (2005) Synthesis and tribological behaviour of metakaolinite-based geopolymer composites. Mater Lett 59:3976–3981

Wang H, Xue M, Cao J (2011) Research on the durability of magnesium phosphate cement. Adv Mater Res 170:1864–1868. doi:10.4028/www.scientific.net/AMR.168-170.1864

Wansom S, Janjaturaphan S, Sinthupinyo S (2010) Characterizing pozzolanic activity of rice husk ash by impedance spectroscopy. Cem Concr Res 40:1714–1722. doi:10.1016/j.cemconres.2010.08.013

WBCSD (2010) End-of-life tyres: a framework for effective management systems. http://www.wbcsd.org/.../Appendices-TiresFrameworkForEffectiveELTManagementSystems-Final.pdf

Weil M, Dombrowski K, Buchwald A (2009) Life-cycle analysis of geopolymers. In: Provis J, Van Deventer J (eds) Geopolymers, structure, processing, properties and applications, ISBN-13: 978 1 84569 449 4. Woodhead Publishing Limited, Abington Hall, Cambridge

Wen S, Chung D (2001) Effect of carbon fiber grade on electrical behavior of carbon fiber reinforced cement. Carbon 39:369–373. doi:10.1016/S0008-6223(00)00127-5

Wen S, Chung D (2005) Strain-sensing characteristics of carbon fiber reinforced cement. ACI Mater J 39:244–248. wings.buffalo.edu/.../Strainsensing%20characteristics%20of%20carbon%20fiber.pdf

Wen S, Chung D (2006) Self sensing of flexural damage and strain in carbon fiber reinforced cement and effect of embedded steel reinforcing bars. Carbon 44:369–373. doi: 10.1016/j.carbon.2005.12.009

Weng L, Sagoe-Crentsil K, Brown T, Song S (2005) Effects of aluminates on the formation of geopolymers. Mater Sci Eng 117:163–168. doi:10.1016/j.mseb.2004.11.008

Wild S, Khatib JM (1997) Portlandite consumption in metakaolin cement paste and mortars. Cem Concr Res 27:137–146. doi:10.1016/S0008-8846(96)00187-1

Winnefeld F, Lothenbach B (2010) Hydration of calcium sulfoaluminate cements—Experimental findings and thermodynamic modeling. Cem Concr Res 40:1239–1247. doi:10.1016/j.cemconres.2009.08.014

Xie Z, Xi Y (2001) Hardening mechanisms of an alkaline-activated class F fly ash. Cem Concr Res 31:1245–1249. doi:10.1016/S0008-8846(01)00571-3

Xu H, Van Deventer J (2000) The geopolymerisation of alumino-silicate minerals. Inter J Mineral Process 59:247–266. doi:10.1016/S0301-7516(99)00074-5

Xu H, Van Deventer J (2002) Geopolymerisation of multiple minerals. Minerals Eng 15:1131–1139. doi:10.1016/S0892-6875(02)00255-8

Yadav I (2008) Laboratory investigations of the properties of the concrete containing recycled plastic aggregates. Master of Engineering in Structural Engineering, Thapar University, Patiala, India

Yang Q, Wu X (1999) Factors influencing properties of phosphate cement-based binder for rapid repair of concrete. Cem Concr Res 29:389–396. doi:10.1016/S0008-8846(98)00230-0

Yang Q, Zhu B, Wu X (2000) Characteristics and durability test of magnesium phosphate cement-based material for rapid repair of concrete. Mater Struct 33:229–234. doi: 10.1007/BF02479332

Yang Q, Zhang S, Wu X (2002) Deicer-scaling resistance of phosphate cement-based binder for rapid repair of concrete. Cem Concr Res 32:165–168. doi:10.1016/S0008-8846(01)00651-2

Yang Z, Shi X, Creighton A, Peterson M (2008) Effect of styrene-butadiene rubber latex on the chloride permeability and microstructure of Portland cement mortar. Constr Build Mater 23:2283–2290. doi:10.1016/j.conbuildmat.2008.11.011

Yellishetty M, Karpe V, Reddy E, Subhash K, Ranjith P (2008) Reuse of iron ore mineral wastes in civil engineering constructions: A case study. Res Conserv Recycl 52:1283–1289. doi: 10.1016/j.resconrec.2008.07.007

Yip C, Van Deventer J (2003) Microanalysis of calcium silicate hydrate gel formed within a geopolymeric binder. J Mater Sci 38:3851–3860. doi:10.1023/A:1025904905176

Yu Q, Sawayama K, Sugita S, Shoya M, Isojima Y (1999) Reaction between rice husk ash and Ca(OH)2 solution and the nature of its product. Cem Concr Res 29:37–43. doi: 10.1016/S0008-8846(98)00172-0

Zain M, Islam M, Mahmud F, Jamil M (2011) Production of rice husk ash for use in concrete as a supplementary cementitious material. Constr Build Mater 25:798–805. doi: 10.1016/j.conbuildmat.2010.07.003

Zerbino R, Giaccio G, Isaia G (2011) Concrete incorporating rice husk ash without processing. Constr Build Mater 25:371–378. doi:10.1016/j.conbuildmat.2010.06.016

Zheng L, Huo S, Yuan Y (2008a) Experimental investigation on dynamic properties of rubberized concrete. Constr Build Mater 22:939–947. doi:10.1016/j.conbuildmat.2007.03.005

Zheng L, Huo X, Yuan Y (2008b) Strength, modulus of elasticity, and brittleness index of rubberized concrete. J Mater Civil Eng 20:692–699. doi:10.1061/(ASCE)0899-1561(2008)20:11(692)

Zhihua P, Dongxu L, Jian Y, Nanry Y (2002) Hydration products of alkali-activated slag red mud cementitious material. Cem Concr Res 32:357–362. doi:10.1016/S0008-8846(01)00683-4

Zhihua P, Dongxu L, Jian Y, Nanry Y (2003) Properties and microstructure of the hardened alkali-activated red mud-slag cimentitious material. Cem Concr Res 33:1437–1441. doi: 10.1016/S0008-8846(03)00093-0

Zornoza E, Garcés P, Payá J, Climent M (2009) Improvement of the chloride ingress resistance of OPC mortars by using spent cracking catalyst. Cem Concr Res 39:126–139

Chapter 6
Masonry Units

6.1 General

While stone masonry walls exist since the beginning of human civilization, the first bricks were based on dried mud and were used for the first time in 8,000 BC in Mesopotamia, an area bordered by the rivers Tigris and Euphrates stretching from Southeast Turkey, Northern Syria and Iraq reaching the Persian Gulf. As to the fired-clay bricks, its use go back to 3,000 BC (Lynch 1994). The ceramic glazed bricks of the Ishtar Gate dating from 500–600 BC show that ceramic bricks reached a level of some sophistication. Although the Roman civilization has left numerous constructions made of stone masonry, they also left several buildings constructed with fired-clay bricks, as it happens in the case of the library of Celsus in Ephesus built in 117 AC. Traditional masonry uses mainly hollow clay bricks and concrete blocks. The environmental impacts of the latter are mostly related to the production of Portland cement (an issue analyzed in Chap. 5) and rather lower when compared to the environmental impacts of fired-clay brick production. According to Reddy and Jagadish (2003) fired-clay brick masonry, has an energy that is almost 300% higher than the energy of concrete block masonry. The environmental impacts caused by the fired-clay brick industry, can be summarized as follows:

- Non-renewable resources consumption
- Energy consumption
- Water consumption
- GHGs emissions
- Waste generation

The majority of the environmental impacts associated with the consumption of nonrenewable resources are less related to the availability of clay, but rather on the reduction of the area that should be available for biodiversity conservation purposes. The need for high temperatures for the production of fired-clay bricks means that this is an industry with an high energy consumption. The energy

F. P. Torgal and S. Jalali, *Eco-efficient Construction and Building Materials*,
DOI: 10.1007/978-0-85729-892-8_6, © Springer-Verlag London Limited 2011

sources cover fuel, natural gas and propane. The use of more efficient equipment, the use of biomass or the use of additives in the composition of the bricks acting as calcination enhancers, contributes to reduce the consumption of fossil fuels. The fired-clay brick industry involves the consumption of high water volumes which, however, are considerably shorter than those required for other industries. The pollutant emissions caused by this industry are made up of particles of sulfur dioxide (SO_2), nitrogen oxide (NOx), carbon monoxide (CO), hydrogen fluoride (HF) and carbon dioxide (CO_2). The wastes generated by this industry are composed mostly of raw and fired-clay pieces. Given its characteristics, this wastes are reused again and incorporated in the production process or may be used as by-products for the production of concrete, as already mentioned in Chap. 5.

6.2 Fired-Clay Bricks with Industrial Wastes

The production of fired-clay bricks with the incorporation of wastes from other industries constitutes a positive way for the ceramic industry to contribute to a more sustainable construction. On one hand there is a reduction of the clay extraction and on the other this avoids the landfill of wastes. Lingling et al. (2005) studied the possibility of replacing large amounts of clay by fly ash. They show that clay-fly ash based bricks need a calcination temperature of almost 1,050°C. This represents between 50°C and 100°C above the traditional calcination temperature. These bricks show a high compressive strength, low water absorption and a high freeze-thaw resistance. Table 6.1 shows that increasing the fly ash/clay ratio leads to a reduction both in compressive strength and in density, as well as an increase in water absorption.

Those authors also mentioned that the use of high volume fly ash leads to a reduction in the plasticity index (Fig. 6.1). Since the mixtures with a plasticity index below six make it difficult to cast bricks by plastic extrusion these means that mixtures with a fly ash/clay ratio above 60% are not recommended. Other authors (Cultrone and Sebastián 2009) also studied the performance of fly ash based bricks confirming that its inclusion helps to decrease the density of the mixture. They reported that the use of fly ash can lead to a color change of the bricks. This may hinder their use in certain exposed applications when the bricks come from different manufacturers. Saboya et al. (2007) studied the replacement of clay by marble waste mud, a by-product of the marble processing industry. Those authors obtained bricks with a high compressive strength concluding that the use of a replacement percentage of 15% and a calcination temperature of 850°C are the most recommendable. El-Mahllawy (2008) studied the feasibility of using granite powder, kaolin and blast furnace slag in the manufacture of fired bricks with high acid resistance. This author recommended the use of a mixture with 50% kaolin, 20% granite powder and 30% blast furnace slag. Ajam et al. (2009) studied the performance of ceramic bricks with partial replacement of clay by phosphogypsum noticing that the addition does not reduce the plasticity of the

Table 6.1 Properties of clay-fly ash bricks (Lingling et al. 2005)

Fly ash/clay ratio (vol%)	Calcination temperature (°C)	Apparent porosity (%)	Water absorption (%)	Density (kg/m^3)	Compressive strength (MPa)
50:50	1,000	35.82	22.18	1,610	50.0
	1,050	30.37	17.62	1,720	98.5
60:40	1,000	39.83	26.94	1,480	25.4
	1,050	36.65	23.62	1,550	39.6
70:30	1,000	40.62	28.08	1,440	21.5
	1,050	39.76	27.54	1,440	27.8
80:20	1,000	42.12	31.26	1,350	14.7
	1,050	39.80	27.86	1,430	25.4

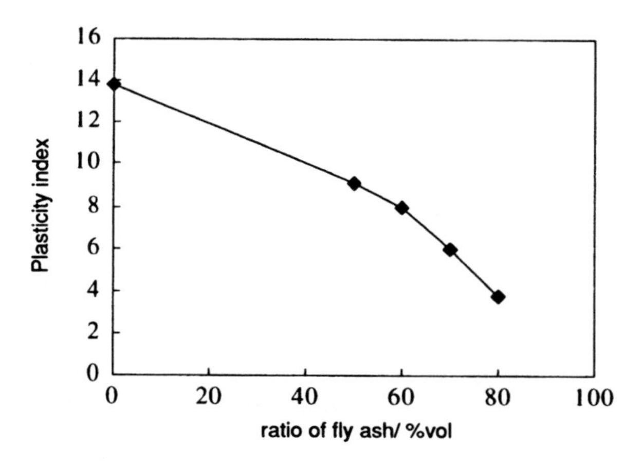

Fig. 6.1 Plasticity indexes of clay and clay-fly ash mixtures (Lingling et al. 2005)

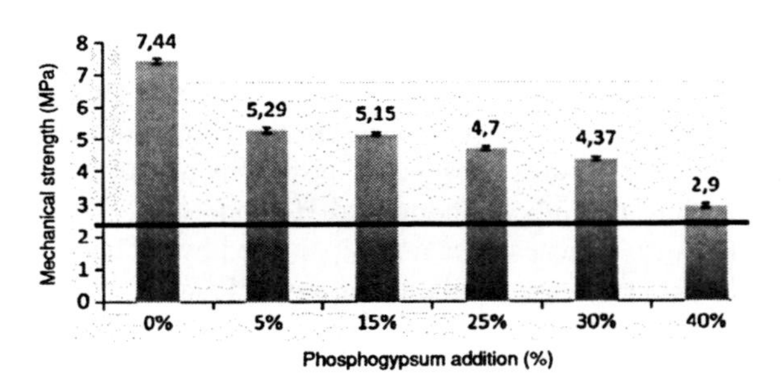

Fig. 6.2 Mechanical strength versus phosphogypsum proportioning (Ajam et al. 2009)

mixture and that the use of substantial amounts of phosphogypsum allows mixtures with enough mechanical strength (Fig. 6.2).

The same authors also noticed that these bricks show a water absorption percentage below the regulatory limits (Table 6.2) and also that the use of phosphogypsum percentages of 5% and 10% lead to a water absorption lower than the one presented by the mixture without phosphogypsum. As to the shrinkage

Table 6.2 Water absorption coefficient of brick samples (%) (Ajam et al. 2009)

$C_{0\%}$	$C_{5\%}$	$C_{15\%}$	$C_{25\%}$	$C_{30\%}$	$C_{40\%}$	Regulatory limits
7.15	5.3	5.7	7.65	11.2	13.4	15

Table 6.3 Shrinkage coefficient of brick samples (%) (Ajam et al. 2009)

$C_{0\%}$	$C_{5\%}$	$C_{15\%}$	$C_{25\%}$	$C_{30\%}$	$C_{40\%}$	Regulatory limits
6.66	6.7	7.2	6.7	7.5	10	8

Table 6.4 Average gaseous emissions (Monteiro et al. 2007)

Gases	Oil wastes (% by weight)	
	0%	10%
SO_2	2 ppm	58 ppm
NO	–	–
CO	5,650 ppm	7,120 ppm
CO_2	3,750 ppm	38,000 ppm
CH_4	–	500 ppm

coefficient (Table 6.3) only the mixture with 40% phosphogypsum show an inadequate behavior.

Monteiro and Vieira (2005) suggest that production of fired-clay bricks can help to solve the problem of oil wastes, thus preventing their disposal. The oil wastes contain water (12.7%), organic matter (33.1%) and some heavy metals. The results show that the use of almost 30% of oil wastes did not alter the density of the fired bricks, nor its water absorption or the linear shrinkage. As to the flexural strength it decreases with increasing percentages of those wastes. Monteiro et al. (2007) also study the use of oil wastes in fired bricks; however they produced the bricks in an industrial facility while other studies were conducted in laboratory using small specimens. These authors show that is possible to produce fired bricks containing oil wastes as long as its percentage does not exceed 5%. They also mentioned that the leaching tests are within the Brazilian thresholds; nevertheless the firing process generates substantial hazardous gaseous emissions (Table 6.4).

More recently Pinheiro and Holanda (2009) confirm that the incorporation of 30% of oil wastes does not impair the physical and mechanical properties of fired-clay bricks. They point out that several authors used different types of waste oil but unfortunately they do not disclose any comment on gaseous emissions. Mekki et al. (2008) studied the possibility of incorporation of olive mill waste water in the fired brick-making process. These wastes have a high organic content and phenols that are toxic and represent an environmental problem. The results showed that the production of fired bricks from the mixture of clay and olive mill waste water allows for a final product with mechanical characteristics identical to bricks without this addition. The new bricks show a 10% increase in shrinkage and a 12% increase in water absorption. The same authors also show that the new bricks can be fired at 880°C instead of the traditional 920°C firing temperature which allows

Table 6.5 Properties of paper processing residues fired-clay bricks (Sutcu and Akkurt 2009)

Properties	Percentage of paper processing residues by weight			
	0	10	20	30
Water absorption (%)	16.7	23.9	31.9	40.4
Compressive strength (MPa)	39.2	15.7	7.5	4.9
Thermal conductivity (W/mK)	0.83	0.59	0.48	0.42

Table 6.6 Total concentrations of heavy metals in raw river sediments in mg/kg on dry material (Samara et al. 2009)

Element	Cadmium	Chromium	Copper	Lead	Zinc
Raw sediment	12.8	413	150.7	1,373	5,032
Level N1	1.2	90	45	100	276
Level N2	2.4	180	90	200	552

for a reduction in the energy consumption. Identical results were obtained by De La Casa et al. (2009) which showed that the reuse of olive mill waste water allows the production of fired bricks with physical and mechanical characteristics similar to traditional fired bricks with the advantage of allowing for energy savings between 2.4% and 7.3%. Cruz (2000) analyzed the performance of fired-clay bricks containing waste sawdust, polystyrene and perlite, mentioning that the new bricks have an increased thermal and acoustic performance. The technique of reducing the density of fired-clay bricks with organic additions takes advantage on the fact that during the firing stage the combustion of the organic matter leads to the formation of micro-pores. This technique has been used by several authors (Kohler 2002; Demir et al. 2005; Demir 2006; Ducman and Kopar 2007). More recently Demir (2008) studied the feasibility of using several organic wastes (sawdust, tobacco residues, grass) to enhance pore formation in fired-clay bricks. The results show that pore formers are not associated with extrusion problems up to 5% weight. A residue addition of 10% weight was found to be unsuitable because of low plasticity and excessive drying shrinkage. Sutcu and Akkurt (2009) used paper processing residues as pore forming agents in fired-clay bricks obtaining new bricks with enhance thermal conductivity (W/m K), high water absorption and adequate compressive strength (Table 6.5).

Samara et al. (2009) studied the use of river sediments in fired-clay bricks. This sediments come from the dredging of river beds that receive effluents from highly polluting industries (coal, iron, steel, glass, chemicals), thus having a high toxic content (Table 6.6).

The levels N1 and N2 are set by the French regulations as toxicity thresholds. Below level N1, the potential impact is regarded, as neutral or negligible. Between levels N1 and N2, further investigations may prove necessary. Beyond N2 level, additional investigations are generally necessary. Since the raw sediment exceeds the level N2 they have been treated with the Novosol® process developed and patented by the Solvay Company. This process encompasses two different phases. A phosphatation phase in which raw sediments are mixed with phosphoric acid

Table 6.7 Results of the leaching test undertaken on brick specimens in accordance with the French Standard AFNOR, XP X31-210 (Samara et al. 2009)

Element	Sediment-amended brick pH 8.9	Standard brick pH 7.6	Limit values for waste acceptable as inert L/S = 10 (l/kg)	Limit values for waste acceptable as non-hazardous L/S = 10 (l/kg)
Cd	<0.02	<0.02	0.04	1
Cu	<0.03	<0.03	2	50
Zn	0.053	0.177	4	50
Ni	<0.07	<0.07	0.4	10
Pb	<0.2	<0.2	0.5	10

Table 6.8 Concentration of heavy metals in the leachates of samples, leached with acetic acid in mg/kg on dry material, according to the American Standard TLCP-USEPA (Samara et al. 2009)

Element	Sediment-amended brick pH 4.92	Standard brick pH 7.6	Regulated TLCP limit
Cd	<0.04	<0.04	1.00
Cu	0.1	0.2	15
Zn	3.7	3.3	25.00
Ni	<0.14	<0.67	–
Pb	<0.4	<0.4	5

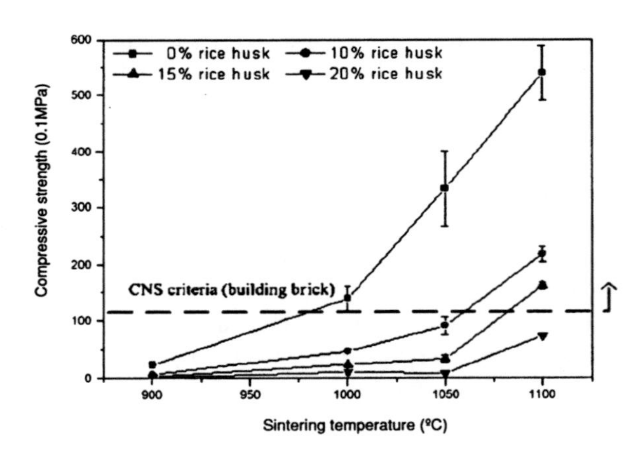

Fig. 6.3 Sintering temperature effect on the compressive strength (Chiang et al. 2009)

H_3PO_4 in the presence of calcite, leading to the formation of calcium phosphates minerals. The second phase implies the calcination of the phosphated sediments at ≥650°C. The treated sediments consisting of an odorless fine powder that were used in fired-clay bricks. The results show that bricks with 15% wastes have increased compressive strength (63%), lower water absorption (13%) and lower porosity (10%). Tables 6.7 and 6.8 shows the leaching performance of the new bricks when using respectively distilled water and acetic acid. The results are within the legal thresholds.

Table 6.9 Chemical composition and heavy metals in TFT-LCD wastes (Lin 2007)

SiO_2	Na_2O	Cu	Zn	Pb	Cr
64%	0.3%	0.27 (mg/kg)	0.23 (mg/kg)	0.65 (mg/kg)	0.18 (mg/kg)

Chiang et al. (2009) studied the reuse of rice husk ash and water treatment sludge to produce light bricks. The results show that the achievement of a minimum regulatory 10 MPa compressive strength implies the use of a calcination temperature of 1,100°C and the use of a rice husk ash percentage below 15% (Fig. 6.3). Water absorption results show that increasing the percentage of rice husk ash leads to high water absorption, which can be reduced by the use of a high sintering temperature.

Lin (2007) studied inert wastes from thin film transistor-liquid crystal display (TFT-LCD) optical waste glass (TVs and computers) incorporated in fired-clay bricks. Estimates about the amount of such waste are around 25,000 m^3/year of PC and TV glass per million people in European countries (Hermans et al. 2001). This represents almost 19 millions of m^3/year. These wastes are composed mostly of glass with some heavy metals (Table 6.9).

The environmental performance of these bricks was examined with the standard TLCP-EPA and all the compositions including those containing 40% wastes met the regulatory limits. The bricks show low water absorption and a high compressive strength both dependent on the firing temperature. The results show that the mixtures with 30% wastes lead to the maximum compressive strength. The reuse of TFT-LCD wastes avoids disposal costs (40 €/ton) and also the cost of raw clay (10 €/ton). Dondi et al. (2009) also studied the inertization of this kind of wastes in fired-clay bricks and roof tiles suggesting the use of only 2%, because higher percentages may be responsible for a plasticity reduction generating extrusion problems, but also for reductions in the compressive strength. These authors used the leaching standard DIN 38414-S4 for the assessment of the environmental performance of the bricks containing TFT-LCD wastes, observing that the metals concentration in the eluates is very low. Loryuenyong et al. (2009) study the reuse of waste glass from structural glass walls in fired-clay bricks. The use of as much as 30% weight waste glass lead to a compressive strength increase of the bricks up to 41 MPa and a water absorption decrease as low as 3%. The use of higher percentages of waste glass lead to a severe decrease in compressive strength and a high water absorption.

6.3 Unfired Units

The use of unfired masonry units allows for low embodied energy units. Unfired-clay units consist of raw clay mixed with sand, compressed and artificially air-dried during one or two-days before being used in construction. The thermal conductivity of unfired-clay bricks follows a linear function related to its density, as it happens for fired-clay bricks (Oti et al. 2010). Usually these units are use to

Table 6.10 Physical and mechanical properties of blocks made with limestone powder wastes and wood wastes (Turgut and Algin 2007)

Mix	Compressive strength (MPa)	Flexural strength (MPa)	Density (g/cm^3)	Water absorption (%)
Ref	24.9	3.94	1.88	12.4
Lw-10	16.6	3.75	1.70	13.9
Lw-20	11.0	3.50	1.66	15.1
Lw-30	7.2	3.08	1.51	19.2

built non-load-bearing walls. According to Morton (2006) the embodied energy of an unfired-clay brick house test is about 14% of the value for fired-clay bricks and 24% for lightweight concrete blocks. Masonry blocks based on hydraulic binders also belong to the unfired units category. Kumar (2000, 2002) mentioned the development of (fly ash + lime + phosphogypsum) based blocks, obtaining a final product with a density between 20% to 40% lower than the fired-clay bricks, but with a compressive strength in the range of 4 to 12 MPa, enough to built masonry walls with a high resistance to aggressive environments. The mixture reproduces the characteristics of a hydraulic binder, the silica in the fly ash reacts with calcium hydroxide to produce calcium silicate hydrates. As to the aluminum in conjunction with calcium hydroxide reacts with gypsum to form calcium tris-sulfoaluminate hydrated. Turgut and Algin (2007) studied the use of limestone powder wastes and wood wastes (10, 20 and 30%), together with small amounts of cement (approx. 10% by mass) in the manufacture of masonry blocks (Table 6.10).

The results show that using a percentage of 30% wood wastes, is responsible for a high reduction of the compressive strength. Still the blocks meet minimum regulatory requirements for materials meant to structural applications, as defined in the BS 6073-1:1981 (Precast concrete masonry units). It is also clear that increasing the percentage of wood wastes leads to increased water absorption and a decrease in the density of the concrete blocks. The same authors (Algin and Turgut 2008) also studied the reuse of limestone wastes and cotton wastes in the production of concrete blocks ($W/C = 0.3$) containing limestone powder and glass wastes (10% to 30%). The results show that increasing the volume of glass wastes means that the compressive strength rises slightly from 27.5 to 30.1 MPa. At the same time the flexural strength increases from 4.15 to 7.76 MPa and the modulus of elasticity increases from 12 to 19 GPa. The results also show that the water absorption remains almost unchanged at about 12%, and that increasing the glass wastes leads to a considerable increase in the freeze-thaw resistance (Fig. 6.4).

Chindaprasirt and Pimraksa (2008) studied the manufacture of blocks based on lime and fly ash (10% + 90%) using an autoclave process (130°C and 0.14 MPa) during 4 h. The fly ash particles were previously submitted to a granulation process that causes a substantial increase in its pozzolanic reactivity because it contributes to an increase of the inter-particle contact. The granulation is obtained by inducing the formation of a water film around the particules. These blocks present a compressive strength between 47 and 62 MPa and a water absorption between 16% and 19%. Pimraksa and Chindaprasirt (2009) used the same

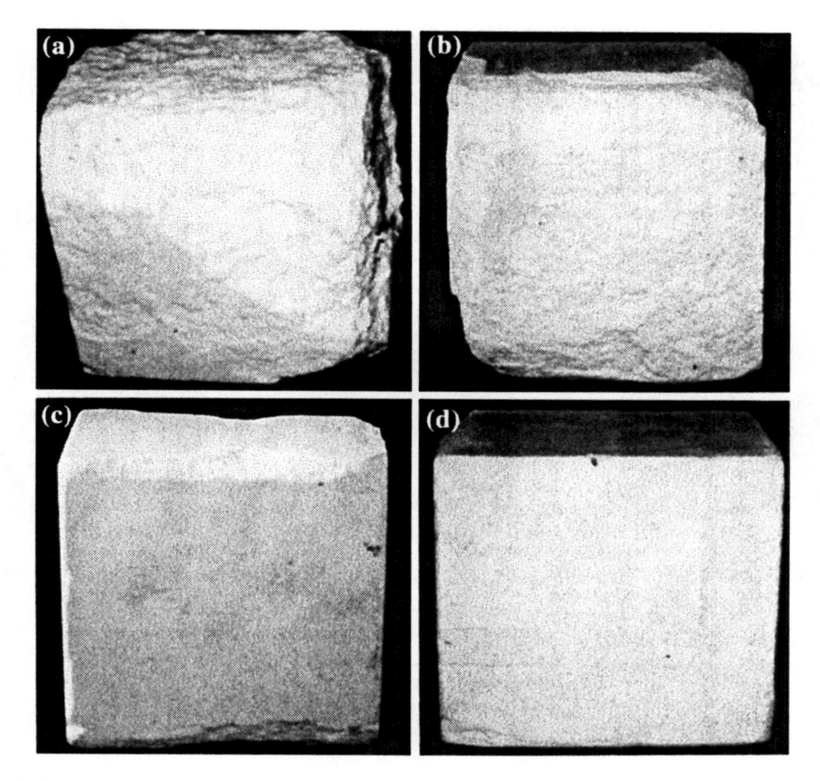

Fig. 6.4 Specimens after 50 freeze-thaw cycles testing: **a** mix without waste glass; **b** mix with 10% waste glass; **c** mix with 20% waste glass; **d** mix with 30% waste glass (Turgut 2008)

autoclave conditions to produce blocks made of diatomaceous earth, lime and gypsum (80% + 15% + 5%) with high compressive strength (14.5 MPa) and low density (880 kg/m^3). Some blocks were made using diatomaceous earth calcined at 500°C showing an increase in the compressive strength (17.5 MPa) and a decrease in their density (730 kg/m^3).

6.4 Shape Optimization

Recent investigations have been carried in order to optimize the shape of masonry units for enhanced thermal and acoustical performance. Dias et al. (2008) present results about the development of highly perforated fired-clay units designated cBloco containing wood wastes as pore formers that allow the construction of single-leaf walls (Fig. 6.5). Table 6.11 presents some of the characteristics of the cBloco unit.

Other authors (Del Coz Diaz et al. 2008, 2011) studied the shape optimization of concrete masonry units in order to reduce its mass and increase its thermal

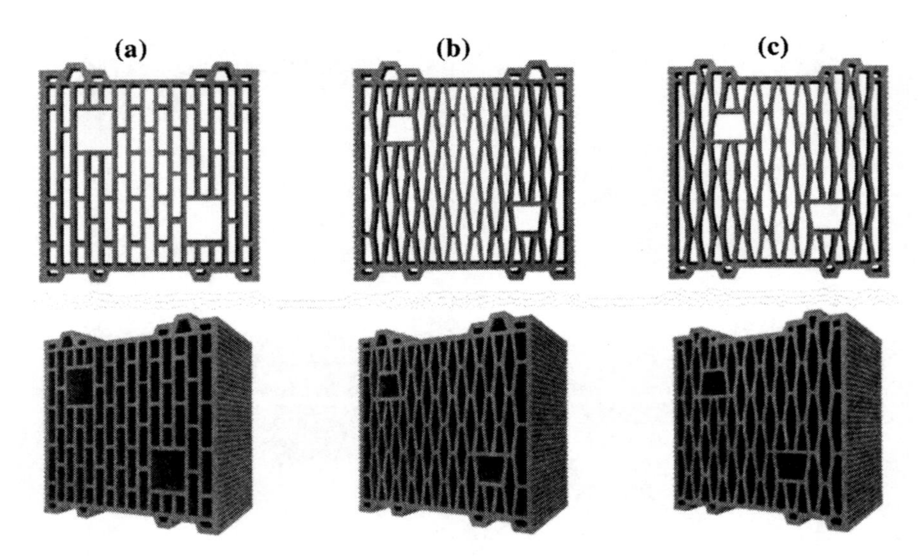

Fig. 6.5 cBloco 30 × 30 × 19 unit: **a** Rectagles; **b** Lozenges; **c** Rice grain (Dias et al. 2008)

Table 6.11 Characteristics of the cBloco unit (Dias et al. 2008)

Characteristics	Value
Dimensions (mm)	300 × 300 × 200
Compressive strengh (MPa)	13
Voids (%)	55
Mass (kg)	14
Real density (kg/m^3)	1,850
Apparent density (kg/m^3)	750
Thermal conductivity-λ (W/mK)	0.50
U-value of the c-Bloco unit (W/m^2 K)	0.60
Acoustic resistance Rw (dB)	44

conductivity. Sousa et al. (2011) studied the shape optimization of lightweight concrete masonry units using a genetic algorithm. The new blocks make it possible to built single walls with a U-value of 0.50 W/m^2 K.

6.5 Conclusions

Traditional masonry units (fired-clay bricks or concrete blocks) without an improved performance in terms of thermal and acoustical insulation are a symbol of a low technology past very far from the demands of eco-efficient construction. The best commercially available solutions for fired-clay bricks and lightweight concrete blocks allow to built single masonry walls with high thermal performance ($U < 0{,}6$ W/m^{2}°C). Therefore, the eco-efficient choice between these two masonry units will be made in terms of its global environmental impact. However, taking into account the low embodied energy of concrete blocks its expected that in the future this material will gain a higher market share. An increase in the use of unfired-clay bricks will also occur. The reuse of wastes from other industries will increase the eco-efficiency of masonry units.

References

Ajam L, Ouezdou M, Felfoul H, Mensi R (2009) Characterization of Tunisian phosphogypsum and its valorization in clay bricks. Constr Build Mater 23:3240–3247. doi:10.1016/j.conbuildmat.2009.05.009

Algin H, Turgut P (2008) Cotton and limestone powder wastes as brick material. Constr Build Mater 22: 1074–1080. eng.harran.edu.tr/~pturgut/6.pdf

Chiang K, Chou P, Hua C, Chien K, Cheeseman C (2009) Lightweight bricks manufactured from water treatment sludge and rice husks. J Hazard Mater 171:76–82. doi:10.1016/j.jhazmat.2009.05.144

Chindaprasirt P, Pimraksa K (2008) A study of fly ash-lime granule unfired brick. Powder Technol 182:33–41. doi:10.1016/j.powtec.2007.05.001

Cruz J (2000) Ceramic blocks with pore formers for enhanced thermal performance. Master Thesis, LNEC-IST, Lisbon

Cultrone G, Sebastián E (2009) Fly ash addition in clayey materials to improve the quality of solid bricks. Constr Build Mater 23: 1178–1184. www.ugr.es/~grupo179/.../Constr%20Build%20Mat%202009.pdf

De La Casa J, Lorite M, Jiménez J, Castro E (2009) Valorization of waste water from two-phase olive oil extraction in fired clay brick production. J Hazard Mater 169:271–278. doi:10.1016/j.jhazmat.2009.03.095

Del Coz Diaz J, Nieto P, Sierra J, Sanchez I (2008) Non-linear thermal optimization and design improvement of a new internal light concrete multi-holed brick walls by FEM. Appl Therm Eng 28:1090–1100. doi:10.1016/j.applthermaleng.2007.06.023

Del Coz Diaz J, Nieto P, Rabanal F, Martínez-Luengas A (2011) Design and shape optimization of a new type of hollow concrete masonry block using the finite element method. Eng Struct 33:1–9. doi:10.1016/j.engstruct.2010.09.012

Demir I (2006) An investigation on the production of construction brick with processed waste tea. Build Environ 41:1274–1278. doi:10.1016/j.buildenv.2005.05.004

Demir I (2008) Effect of organic residues addition on the technological properties of clay bricks. Waste Manag 28:622–627. doi:10.1016/j.wasman.2007.03.019

Demir I, Baspinar M, Orhan M (2005) Utilization of kraft pulp production residues in clay brick production. Build Environ 40:1533–1537. doi:10.1016/j.buildenv.2004.11.021

Dias A, Sousa H, Lourenço P, Ferraz E, Sousa L, Sousa R, Vasconcelos G, Medeiros P (2008) Development of a sustainable fired-caly brick for sustainable construction. Congress on inovation for sustainable construction CINCOS′08. Centro Habitat, Cúria, Portugal, pp 165–172

Dondi M, Guarini G, Raimondo M, Zanelli C (2009) Recycling PC and TV waste glass in clay bricks and roof tiles. Waste Manag 29:1945–1951. doi:10.1016/j.wasman.2008.12.003

Ducman V, Kopar T (2007) The influence of different waste additions to clay-product mixtures. Mater Technol 41:289–293

El-Mahllawy M (2008) Characteristics of acid resisting bricks made from quarry residues and waste steel slag. Constr Build Mater 22:1887–1896. doi:10.1016/j.conbuildmat.2007.04.007

Hermans J, Peelen J, Bei J (2001) Recycling of the TV glass: profit or doom? Am Ceram Soc Bull 80:51–56

Kohler R (2002) Use of leather residues as pore-forming agents for masonry bricks. Ziegelind Inter 58:30–38. doi:10.1016/j.ceramint.2009.02.027

Kumar S (2000) Fly-ash-lime phosphogypsum cementitious binder: anew trend in bricks. Mater Struct 33:59–64

Kumar S (2002) A perspective study on fly ash-lime-gypsum bricks and hollow blocks for low cost housing development. Constr Build Mater 16:519–525. doi:10.1016/S0950-0618(02)00034-X

Lin K (2007) The effect of heating temperature of thin film transistor-liquid crystal display (TFT-LCD) optical waste glass as a partial substitute partial for clay in eco-brick. J Clean Prod 15:1755–1759. doi:10.1016/j.jclepro.2006.04.002

Lingling X, Wei G, Tao W, Nanru Y (2005) Study on fired bricks with replacing clay by fly ash in high volume ratio. Constr Build Mater 19:243–247. doi:10.1016/j.conbuildmat.2004.05.017

Loryuenyong V, Panyachai T, Kaewsimork K, Siritai C (2009) Effects of recycled glass substitution on the physical and mechanical properties of clay bricks. Waste Manag 29:2717–2721. doi:10.1016/j.wasman.2009.05.015

Lynch G (1994) Brickwork: history, technology and practice. Donhead, London

Mekki H, Anderson M, Benzina M, Ammar E (2008) Valorization of olive mill wastewater by its incorporation in building bricks. J Hazard Mater 158:308–315. doi:10.1016/j.jhazmat.2008.01.104

Monteiro S, Vieira C (2005) Effect of oily waste addition to clay ceramic. Ceram Inter 31:353–358. doi:10.1016/j.ceramint.2004.05.002

Monteiro S, Vieira C, Ribeiro M, Silva F (2007) Red ceramic industrial products incorporated with oily wastes. Constr Build Mater 21:2007–2011. doi:10.1016/j.conbuildmat.2006.05.035

Morton T (2006) Feat of clay. http:www.iom3.org/materialsworld/feature-pdfs

Oti J, Kinuthia J, Bai J (2010) Design thermal values for unfired clay bricks. Mater Des 31:104–112. doi:10.1016/j.matdes.2009.07.011

Pimraksa K, Chindaprasirt P (2009) Lightweight bricks made of diatomaceous earth, lime and gypsum. Ceram Inter 35:471–478. doi:10.1016/j.ceramint.2008.01.013

Pinheiro B, Holanda J (2009) Processing of red ceramics incorporated with encapsulated petroleum waste. J Mater Process Technol 209:5606–5610. doi:10.1016/j.jmatprotec.2009.05.018

Reddy B, Jagadish K (2003) Embodied energy of common and alternative building materials and technologies. Energy Build 35:129–137. profile.iiita.ac.in/.../Green%20Building%20Training%20Programme/Emb%20energy%20materials2.pdf

Saboya F, Xavier G, Alexandre J (2007) The use of the powder marble by-product to enhance the properties of brick ceramic. Constr Build Mater 21:1950–1960. doi:10.1016/j.conbuildmat.2006.05.029

Samara M, Lafhaj Z, Chapiseau C (2009) Valorization of stabilized river sediments in fired clay bricks: factory scale experiment. J Hazard Mater 163:701–710. doi:10.1016/j.jhazmat.2008.07.153

Sousa L, Castro C, Carlos A, Sousa H (2011) Topology optimisation of masonry units from the thermal point of view using a genetic algorithm. Constr Build Mater 25:2254–2262. doi:10.1016/j.conbuildmat.2010.11.010

Sutcu M, Akkurt S (2009) The use of recycled paper processing residues in making porous brick with reduced thermal conductivity. Ceram Inter 35:2625–2631. doi:10.1016/j.ceramint.2009.02.027

Turgut P (2008) Limestone dust and glass powder wastes as new brick material. Mater Struct 41:805–813. doi:10.1617/s11527-007-9284-3

Turgut P, Algin H (2007) Limestone dust and wood sawdust as brick material. Constr Build Mater 42:3399–3403. doi:10.1016/j.buildenv.2006.08.012

Chapter 7
Cement Composites Reinforced with Vegetable Fibres

7.1 General

The use of construction and building materials made from renewable resources is generally regarded as an indispensable option so the construction industry can become more sustainable. That premise can not however be taken in absolute terms since not all situations involving the use of renewable resources like wood or other plant species are exempt from any environmental impact. This is the case of woods with high environmental impact through its transport over long distances or those that use large amounts of fertilizers, pesticides, fungicides or involving the destruction of ecosystems during the growth phase (Swanson and Franklin 1992; Powers 1999; Sample 2006; Burger 2009). One of the worst examples of this kind of ecological disaster can be found in the regions of Sumatra, Borneo, and Malaysia, where millions of square kilometers of rainforest were destroyed for the production of oil from palm three. This option endangered the survival of hundreds of species which include some mammals such as elephants, tigers, rhinos, and orangutans (UNEP 2007). Similar considerations can be made about the destruction of tropical forests to produce timber for industrial use or about the harvest of exotic woods at a rate higher than their natural regeneration. The use of toxic products for the protection of wood products, previously mentioned in Chap. 2, can not be considered a very sustainable option. A part from the aforementioned cases and as long as wood comes from certified forests (Rametsteiner and Simula 2003) it can be said that the resurgence of this material can only be viewed with an environmentally optimism. The use of wood species in the production of cement composites is particularly interesting for the construction industry. Furthermore, due to cancer health risks (Azuma et al. 2009; Kumagai and Kurumatani 2009) the Directive 83/477/EEC and amending Directives 91/382/EEC, 98/24/EC; 2003/18/EC and 2007/30/EC forbid the production of cementitious products based on fibre silicates (asbestos). Mineral fibres are now being replaced by synthetic fibres like polyvinyl alcohol (PVA) and polypropylene to produce fibre–cement products

F. P. Torgal and S. Jalali, *Eco-efficient Construction and Building Materials*,
DOI: 10.1007/978-0-85729-892-8_7,

Fig. 7.1 Production of fibre–cement composites by the Hatschek process (Ikai et al. 2010)

using the Hatscheck process. An industrial process that represents 85% of fibre–cement composites production worldwide (Fig. 7.1).

However, production of PVA and polypropylene needs phenol compounds as antioxidants and amines as ultraviolet stabilizers and other additives acting as flame retardant which is not the right path to obtain more eco-efficient materials. This represents a large opportunity in the field of vegetable fibres cement based materials because they are as stronger as synthetic fibres, cost-effective and above all environmentally friendly. Therefore to promote the use of cementitious building materials reinforced with vegetable fibres could be a way to achieve a more eco-efficient construction. Another interesting possibility related to use of vegetable fibres encompasses the replacement of steel bars in reinforced concrete. Concrete is known for its high compressive strength and low tensile strength. The combined use of regular concrete and steel reinforced bars is needed to overcome that disadvantage leading to a material with good compressive and tensile strengths but also with a long post-crack deformation (strain softening). Unfortunately reinforced concrete has a high permeability that allows water and other aggressive elements to penetrate, leading to carbonation and chloride ion attack resulting in corrosion problems (Glasser et al. 2008; Bentur and Mitchell 2008). Steel rebar corrosion is in fact the main reason for infrastructure deterioration. Gjorv (1994) mentioned a study of Norway OPC bridges indicating that 25% of those built after 1970 presented corrosion problems. Another author (Ferreira 2009) mentioned that 40% of the 600,000 bridges in the USA were affected by corrosion problems and estimated in 50 billion dollars the repairing operations

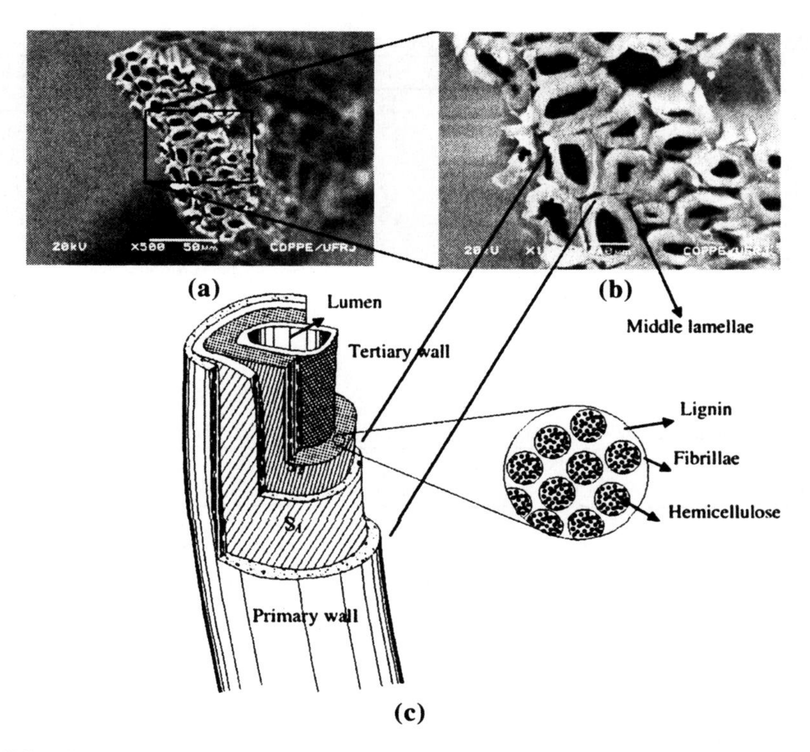

Fig. 7.2 Fibre–cell microstructure: **a** cross-section view showing the fibre–cells, lumens and middle lamellae, **b** magnification of the cross-section, and **c** schematic drawing showing the different layers of an individual fibre–cell (Filho et al. 2009)

cost. Since an average of 200 kg of steel rebar is used for each cubic meter of concrete structure it is clear that the replacement of reinforced steel rebar by vegetable fibres is a major step to achieve a more eco-efficient construction. On the other hand, reinforced steel is a highly expensive material, has high energy consumption and comes from a non renewable resource, while vegetable fibres are available almost all over the world (Brandt 2008).

7.2 Fibre Characteristics and Properties

Vegetable fibres are natural composites with a cellular structure. Different proportions of cellulose, hemi cellulose and lignin constitute the different layers. Cellulose is a polymer containing glucose units. Hemi cellulose is a polymer made of various polysaccharides. As for lignin it is an amorphous and heterogeneous mixture of aromatic polymers and phenyl propane monomers (John et al. 2005; Filho et al. 2009). Figure 7.2 shows the microstructure and a schematic representation of vegetable fibres. Different fibres have different compositions

Table 7.1 Composition of vegetable fibres (Arsene et al. 2007)

Fibre	Lignin (%)	Cellulose (%)	Hemicellulose (%)	Extractives (%)	Ash (%)
Bagasse	21.8	41.7	28.00	4.00	3.50
Banana leaf	24.84	25.65	17.04	9.84	7.02
Banana trunk	15.07	31.48	14.98	4.46	8.65
Coconut coir	46.48	21.46	12.36	8.77	1.05
Coconut tissue	29.7	31.05	19.22	1.74	8.39
Eucalyptus	25.4	41.57	32.56	8.20	0.22
Sisal	11.00	73.11	13.33	1.33	0.33

Table 7.2 Properties of natural and synthetic fibres (Arsene et al. 2007)

Properties	Specific gravity (kg/m^3)	Water absorption (%)	Tensile strength (MPa)	Modulus of elasticity (GPa)
Sisal	1,370	110	347–378	15.2
Coconut	1,177	93.8	95–118	2.8
Bamboo	1,158	145	73–505	10–40
Hemp	1,500	85–105	900	34
Caesar weed	1,409	182	300–500	10–40
Banana	1,031	407	384	20–51
Piassava palm	1,054	34–108	143	5.6
Date palm (Kriker et al., 2005)	1,300–1,450	60–84	70–170	2.5–4
Polypropylene	913	–	250	2.0
PVA F45 (Passuello et al., 2009)	1,300	–	900	23

(Table 7.1) therefore it is expected that their behavior inside a cement matrix could differ between them.

Natural fibres have a high tensile strength and a low modulus of elasticity (Table 7.2). Even so, their tensile performance can stand in a favorable manner with synthetic ones. One of the disadvantages of using natural fibres is that they have a high variation on their properties which could lead to unpredictable concrete properties (Swamy 1990; Li et al. 2006).

Pre-treatment of natural fibres was found to increase fibre reinforced concrete performance. Pulping is one of the fibre treatments that improve fibre adhesion to the cement matrix and also the resistance to the alkaline attack (Savastano et al. 2001b). Pulping can be obtained by a chemical process (kraft) or a mechanical one. Table 7.3 presents some pulping conditions for sisal and banana fibres.

Some chemical treatments lead to a higher mechanical performance than others (Pehanich et al. 2004). The pulping process through mechanical conditions has a lower cost (around half) when compared to the use of chemical conditions and has no need for effluent treatments (Savastano et al. 2001a). Some authors suggest the

Table 7.3 Sisal and banana kraft pulping conditions (Savastano et al. 2003)

Parameter	Sisal	Banana
Active alkali (as Na_2O) (%)	9	10
Sulphidity (as Na_2O) (%)	25	25
Liquor/fibre ratio	5:1	7:1
Temperature (°C)	170	170
Digestion time	~75–120 min temperature cook	~85–120 min temperature cook
Total yield (%w/w)	55.4	45.9
Screened yield (%w/w)	45.5	45.3

use of organofunctional silane coupling agents to reduce the hydrophilic behavior of vegetable fibres (Castellano et al. 2004; Abdelmouleh et al. 2004). But recently Joaquim et al. (2009) compared the performance of cementitious composites reinforced by kraft pulp sisal fibres and by sisal fibres modified by the organosolv process. They found out that the best mechanical performance was achieved by the composites with kraft pulp fibres. Arsene et al. (2007) suggests that using a pyrolisis process can increase the fiber strength by a factor of three.

7.3 Matrix Characteristics

Savastano et al. (2000) mentioned that acid compounds released from natural fibres reduce the setting time of the cement matrix. Fibre sugar components, hemi cellulose and lignin can contribute to prevent cement hydration (Bilba et al. 2003; Stancato et al. 2005). According to Sedan et al. (2008), fibre inclusion can reduce the delay of setting by 45 min. The explanation relies on the fact that pectin (a fibre component) can fix calcium preventing the formation of CSH structures. The interfacial transition zone (ITZ) between concrete and natural fibres is porous, cracked and rich in calcium hydroxide crystals (Savastano and Agopyan 1999). Those authors reported a 200 μm thick at 180 days. On the contrary other authors (Savastano et al. 2005a) reported that using vacuum dewatering and high pressure applied after molding led to a dense ITZ (Fig. 7.3a) also reporting that some fibers were free of hydration products Fig. 7.3b).

The use of water-repellents also leads to a good bond between natural fibre and concrete (Ghavami 1995). The mechanical treatment of the fibres also improves the bonding between the fibre and cement (Coutts 2005). Alkaline treatment of fibres improves their strength and also fibre-matrix adhesion (Sedan et al. 2008). Tonoli et al. (2009) compared cement composites with vegetable fibres previously submitted to surface modification with methacryloxypropyltri-methoxysilane (MPTS) and aminopropyltri-ethoxysilane (APTS). The results of cement composites with fibres modified by MPTS showing fibres free from cement hydration products while APTS based fibres presented accelerated mineralization which leads to higher embrittlement behavior of cement composites (Tonoli et al. 2009).

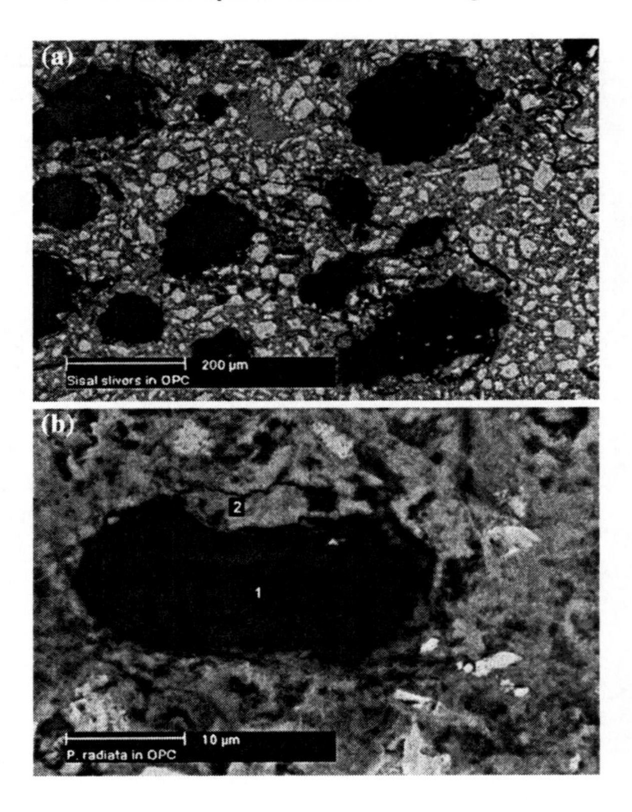

Fig. 7.3 **a** BSE image of sisal fibres in cement matrix with dense ITZ, **b** EDS analysis on *Pinus radiata* fibre lumen (*spot 1*) revealed that no mineralization due to the presence of hydration products was detected (Savastano et al. 2005a)

7.4 Properties of Cement Composites

7.4.1 Using Small Vegetable Fibres

Some authors found out that the use of a 0.2% volume fraction of 25 mm sisal fibres leads to free plastic shrinkage reduction. The combined use of coconut and sisal short fibres seem to have delayed the restrained plastic shrinkage, thus controlling shrinkage and controlling crack development at early ages (Filho et al. 2005). As for the mechanical performance of natural fibre concrete, Al-Oraimi and Seibi (1995) reported that using a low percentage of natural fibres improved the mechanical properties and the impact resistance of concrete, having similar performance when compared to synthetic fibre concrete. Other authors reported that fibre inclusion increases impact resistance by 3 to 18 times higher than when no fibres were used (Ramakrishna and Sundararajan 2005). The use of small volumes (0.6 to 0.8%) of *Arenga Pinata* fibres show an increase in the toughness characteristics of cement-based composites (Razak and Ferdiansyah 2005). As for Reis (2006), their studies showed that the mechanical performance of fibre concrete depends on the type of the fibre. He found that coconut and sugar cane bagasse fibre increases concrete fracture toughness, but banana pseudo stem fibre does not. The use of coconut fibres showed even better flexural strength than does synthetic fibre (glass and carbon) concrete. Silva and Rodrigues (2007) studied the addition of sisal fibres to concrete and reported that the compressive strength was lower than concrete samples without the fibres. The explanation for that behavior seems to be related to concrete workability. Savastano et al. (2009) compared the mechanical performance of cement composites reinforced with sisal, banana and eucalyptus fibres. Sisal and banana fibres with higher lengths (1.65 mm and 1.95 mm) than eucalyptus (0.66 mm) showed a more stable fracture behavior which confirms that fibre length influences the process by which the load transfers from the matrix to the fibres. Silva et al. (2010) tested cement composites reinforced by long sisal fibres placed at the full length of a steel mold in five layers (mortar/fibres/mortar). These composites reach ultimate strengths of 12 and 25 MPa under tension and bending loads. The vegetable type also influences the performance of fibre reinforced cement composites (Tonoli et al. 2010a), being that eucalyptus-based ones present improved mechanical performance after 200 ageing cycles when compared to pinus based ones. The explanation relates to a better distribution of vegetable particles in the cement matrix.

7.4.2 Using Bamboo Rebars

Khare (2005) tested several concrete beams made with stirrups and rebar bamboo and reported that this material has the potential to be used as a substitute for steel reinforcement (Fig. 7.4).

This author reported that the ultimate load capacity of bamboo reinforced concrete beams was about 35% of the equivalent reinforced-steel concrete beams. The strength reduction was due to the low adhesion between the cement matrix and the bamboo rebars. Júnior et al. (2005) mentioned just 25% of the equivalent reinforced-steel concrete beams ultimate load capacity. Analysis of adhesion between cement and bamboo by pull-off tests shows that bamboo/cement have much lower adhesion than steel rebar/cement and that adhesion results are influence by node presence (Jung 2006).

These authors suggest that bamboo rebar should previously be submitted to a thermal treatment to improve bond strength. According to Mesquita et al. (2006), the bond strength of bamboo is 70% of smooth steel bond strength when a 35 MPa concrete is used. However the bond strength of bamboo is almost 90% of smooth steel bond strength when a 15 MPa concrete is used. These authors analyzed the effect of using artificial pins (two of bamboo and two of steel) in bamboo splints, noticing they led to a bond strength of bamboo higher than smooth steel. Ferreira (2007) also studied the effect of artificial pins (Fig. 7.5) in the bond strength of bamboo rebar using pull-out tests.

The results show that the use of just one pin is insufficient to increase bamboo bond strength (Table 7.4). In the same work this author study several 20 MPa concrete beams reinforced with bamboo rebar's (2×1 cm^2), and steel stirrups mentioning an acceptable structural behavior.

Fig. 7.4 Concrete beam reinforced with bamboo rebars and bamboo stirrups: **a** finished bamboo reinforcement, **b** finished reinforcement in the form, **c** test set-up (Khare 2005)

Fig. 7.5 Several bamboo rebars: The first with hole, the second with bamboo pin and the third with steel pin (Ferreira 2007)

Table 7.4 Bond strength using pull-out tests (Ferreira 2007)

Rebar type	Bond strength (MPa)
Bamboo	0.81
Bamboo with epoxy resin	0.32
Bamboo with one bamboo pin	0.82
Bamboo with one steel pin	0.69
Bamboo with hole	1.10
Rough steel	6.87
Smooth steel	1.33

7.5 Durability

Durability of natural fibre reinforced concrete is related to the ability to resist both external (temperature and humidity variations, sulfate or chloride attack, etc.) and internal damages (compatibility between fibres and cement matrix, volumetric changes, etc.). The degradation of natural fibres immersed in Portland cement is due to the high alkaline environment that dissolves the lignin and hemicellulose phases, thus weakening the fibre structure (Gram 1983). Gram was the first author to study the durability of sisal and coir fibre reinforced concrete. The fibre degradation was evaluated by exposing them to alkaline solutions and then measuring the variations in tensile strength. This author reported a deleterious effect of Ca^{2+} elements on fibre degradation. He also stated that fibres were able to preserve their flexibility and strength in areas with carbonated concrete with a pH of nine or less. Filho et al. (2000) also investigated the durability of sisal and coconut fibres when immersed in alkaline solutions. Sisal and coconut fibres conditioned in a sodium hydroxide solution retained respectively, 72.7% and 60.9% of their initial strength after 420 days. As for the immersion of the fibres in a calcium hydroxide solution, it was noticed that the original strength was completely lost after 300 days. According to those authors, the explanation for the higher attack by $Ca(OH)_2$ can be related to a crystallization of lime in the fibres' pores. Ramakrishna and Sundararajan (2005a) also reported the degradation of vegetable fibres when exposed to alkaline media. Other authors studied date palm-reinforced concrete, reporting low durability performance that is related to fibre degradation when immersed in alkaline solutions (Kriker et al. 2008). Ghavami (2005) reported the case of a 15-years-old bamboo-reinforced concrete beam without any deterioration signs. Lima et al. (2008) studied the variations of tensile strength, and Young's modulus of bamboo reinforced concrete exposed to wetting and drying cycles, reporting insignificant changes, thus confirming its durability. The capacity of vegetable fibres to absorb water is another path to decreasing the durability of fibre reinforced concrete. Water absorption leads to volume changes that can induce concrete cracks (Ghavami 2005; Agopyan et al. 2005). In order to improve the durability of fibre reinforced concrete, the two following paths could be used.

7.5.1 Matrix Modification

Using low alkaline concrete and adding pozzolanic by-products such as rice husk ash, blast furnace slag, or fly ashes to Portland cement (Gutiérrez et al. 2005; Agopyan et al. 2005; Savastano et al. 2005a). Results show that the use of ternary blends containing slag/metakaolin and silica fume are effective in preventing fibre degradation (Mohr et al. 2007). But in some cases the low alkalinity is not enough to prevent lignin from being decomposed (John et al. 2005). Other authors reported that fast carbonation can induce lower alkalinity (Agopyan et al. 2005). These results are confirmed by other authors that used artificial carbonation in order to obtain $CaCO_3$ from $Ca(OH)_2$ leading to increased strength and reduced water absorption (Tonoli et al. 2010b). The use of cement-based polymers can also contribute to an increased durability (Pimentel et al. 2006). D'Almeida et al. (2009) used blends where 50% of Portland cement was replaced by metakaolin to produce a matrix totally free of calcium hydroxide in order to prevent migration of calcium hydroxide to the fibre lumen, middle lamella and cell walls, thus avoiding an embrittlement behavior.

7.5.2 Fibre Modification

Coating natural fibres to avoid water absorption and free alkalis. Use waterrepellent agents or fibre impregnation with sodium silicate, sodium sulphite, or magnesium sulphate. Ghavami (1995) reported that using a water-repellent in bamboo fibres allowed only 4% water absorption. The use of organic compounds such as vegetable oils reduced the embrittlement process, but not completely (Filho et al. 2003). Recent findings report that silane coating of fibres is a good way to improve the durability of vegetable fibre reinforced concrete (Bilba and Arsene 2008). Other authors mentioned that using pulped fibres may improve the durability performance (Savastano et al. 2001b). Some authors even reported that the fibre extraction process can prevent durability reductions (Juárez et al. 2007). The use of compression and temperature (120°C, 160°C and 200°C) leads to an increase of fibre stiffness and a decrease on the fibre water absorption (Motta et al. 2009).

7.6 Conclusions

The replacement of asbestos and synthetic fibres by vegetable fibres in the manufacture of cementitious composites could contribute to a higher eco-efficiency of the construction industry. The same happens with the replacement of steel bars by bamboo rebars. Further investigations about vegetable fibre reinforced concrete are needed in order to clarify several aspects that current knowledge does not.

The available literature data is mostly related to the mechanical behavior of vegetable fibre reinforced concrete. For instance, only recently has the delaying effect of vegetable fibre inclusion received the proper attention. Since the main reason for vegetable fibre degradation relates to alkaline attack, much more research is needed about the chemical interactions between the cement matrix and the vegetable fibres. The right treatments to improve fibre and cement matrix compatibility are still to be found. The same could be said about the variation on the fibre properties, therefore control quality methods are needed in order to ensure minimal variations on the properties of vegetable fibres. Durability related issues also deserve more research efforts. For concrete with bamboo rebars investigations to improve the adhesion to the cement paste are still needed.

References

Abdelmouleh M, Boufi S, Belgacem M, Duarte A, Salah A, Gandini A (2004) Modification of cellulosic fibres with funcionalised silanes: development of surface properties. Inter J Adhes Adhes 24:43–54

Agopyan V, Savastano H, John V, Cincotto M (2005) Developments on vegetable fibre-cement based materials in São Paulo, Brazil: an overview. Cem Concr Compos 27:527–536. doi: 10.1016/j.cemconcomp.2004.09.004

Al-Oraimi S, Seibi A (1995) Mechanical characterization and impact behavior of concrete reinforced with natural fibres. Compos Struct 32:165–171. doi:10.1016/0263-8223(95)00043-7

Arsene M, Okwo A, Bilba K, Soboyejo A, Soboyejo W (2007) Chemically and thermally treated vegetable fibers for reinforcement of cement-based composites. Mater Manufact Process 22:214–227. doi:10.1080/10426910601063386

Arsène M-A, Savastano H Jr, Allameh S, Ghavami K, Soboyejo W (2003) Cementitious composites reinforced with vegetable fibers. In: Proceedings of the First Interamerican conference on non-conventional materials and technologies in the Eco-construction and Infrastructure, IAC- NOCMAT 2003, Joao-Pessoa, Brazil

Azuma K, Uchiyama I, Chiba Y, Okumura J (2009) Mesothelioma risk and environmental exposure to asbestos: Past and future trends in Japan. Int J Occup Environ Health 15:166–172

Bentur A, Mitchell D (2008) Material performance lessons. Cem Concr Res 38:259–272. doi: 10.1016/j.cemconres.2007.09.009

Bilba K, Arsene M (2008) Silane treatment of bagasse fiber for reinforcement of cementitious composites. Compos A 39:1488–1495. doi:10.1016/j.compositesa.2008.05.013

Bilba K, Arsene M, Ouensanga A (2003) Sugar cane bagasse fibre reinforced cement composites. Part I. Influence of the botanical components of bagasse on the setting of bagasse/cement composite. Cem Concr Compos 25:91–96. doi:10.1016/S0958-9465(02)00003-3

Brandt A (2008) Fibre reinforced cement-based (FRC) composites after over 40 years of development in building and civil engineering. Compos Struct 86:3–9. doi:10.1016/j.compstruct.2008.03.006

Burger J (2009) Management effects on growth, production and sustainability of managed forest ecosystems: past trends and future directions. For Ecol Manag 17:1335–2346. doi:10.1016/j.foreco.2009.03.015

Castellano M, Gandini A, Fabbri P, Belgacem M (2004) Modification of cellulose fibres with organosilanes: under what conditions does coupling occur? J Coll Interface Sci 273:505–511. doi:10.1016/j.jcis.2003.09.044

Coutts R (2005) A review of Australian research into natural fibre cement composites. Cem Concr Compos 27:518–526. doi:10.1016/j.cemconcomp.2004.09.003

D'Almeida A, Filho J, Filho R (2009) Use of curaua fibers as reinforcement in cement composites. Chem Engin Trans 17:1717–1722. www.aidic.it/icheap9/webpapers/146D'Almeida.pdf

Ferreira G (2007) Vigas de concreto armadas com taliscas de bamboo Dendrocalamus Giganteus. Ph.D. Thesis, UNICAMP, Brazil

Ferreira RM (2009) Service-life Design of Concrete Structures in Marine Environments: A probabilistic based approach. VDM Verlag Dr. Muller Aktiengesellschaft & Co. KG

Filho R, Scrivener K, England G, Ghavami K (2000) Durability of alkali-sensitive sisal and coconuts fibres in cement mortar composites. Cem Concr Compos 22:127–143. doi:10.1016/S0958-9465(99)00039-6

Filho R, Ghavami K, England G, Scrivener K (2003) Development of vegetable fibre-mortar composites of improved durability. Cem Concr Compos 25:185–196. doi:10.1016/S0958-9465(02)00018-5

Filho R, Ghavami K, Sanjuán M, England G (2005) Free, restrained and drying shrinkage of cement mortar composites reinforced with vegetable fibres. Cem Concr Compos 27:537–546. doi:10.1016/j.cemconcomp.2004.09.005

Filho RD, Silva FS, Fairbarn E, Filho JA (2009) Durability of compression molded sisal fiber reinforced mortar laminates. Constr Buid Mater 23:2409–2420. doi:10.1016/j.conbuildmat.2008.10.012

Ghavami K (1995) Ultimate load behaviour of bamboo-reinforced lightweight concrete beams. Cem Concr Compos 17:281–288. doi:10.1016/0958-9465(95)00018-8

Ghavami K (2005) Bamboo as reinforcement in structure concrete elements. Cem Concr Compos 27:637–649. doi:10.1016/j.cemconcomp.2004.06.002

Gjorv O (1994) Steel corrosion in concrete structures exposed to Norwegian marine environment. ACI Concr Int 35–39

Glasser F, Marchand J, Samson E (2008) Durability of concrete. Degradation phenomena involving detrimental chemical reactions. Cem Concr Res 38:226–246. doi:10.1016/j.cemconres.2007.09.015

Gram H (1983) Durability of natural fibres in concrete. Swedish Cement and Concrete Research Institute, Stockolm

Gutiérrez R, Díaz L, Delvasto S (2005) Effect of pozzolans on the performance of fiber-reinforced mortars. Cem Concr Compos 27:593–598. doi:10.1016/j.cemconcomp.2004.09.010

Ikai S, Reicher J, Rodrigues A, Zampieri V (2010) Asbestos-free technology with new high toughness polypropylene (PP) fibers in air-cured Hatschek process. Constr Build Mater 24:171–180. doi:10.1016/j.conbuildmat.2009.06.019

Joaquim A, Tonoli G, Santos S, Savastano H (2009) Sisal organosolv pulp as reinforcement for cement based composites. Mater Res 12:305–314. doi:10.1590/S1516-14392009000300010

John V, Cincotto M, Sjotrom C, Agopyan V, Oliveira C (2005) Durability of slag mortar reinforced with coconut fibre. Cem Concr Compos 27:565–574. doi:10.1016/j.cemconcomp.2004.09.007

Juárez C, Durán A, Valdez P, Fajardo G (2007) Performance of *Agave lechuguilla* natural fiber in Portland cement composites exposed to severe environment conditions. Build Environ 42:1151–1157. doi:10.1016/j.buildenv.2005.12.005

Jung Y (2006) Investigation of bamboo as reinforcement in concrete. Master of Science in Civil and Environment Engineering. University of Texas,

Júnior H, Mesquita L, Fabro G, Willrich F, Czarnieski C (2005) Concrete beams reinforced with bamboo *Dendrocalamus giganteus*. I: Experimental analysis. R Bras Eng Agr Ambient 9:642–651

Khare L (2005) Performance evaluation of bamboo reinforced concrete beams. Master of Science in Civil Engineering. University of Texas

Kriker A, Debicki G, Bali A, Khenfer M, Chabannet M (2005) Mechanical properties of date palm fibres and concrete reinforced with date palm fibres in hot dry climates. Cem Concr Compos 27:554–648. doi:10.1016/j.cemconcomp.2004.09.015

Kriker A, Bali A, Debicki G, Bouziane M, Chabannet M (2008) Durability of date palm fibres and their use as reinforcement in hot dry climates. Cem Concr Compos 30:639–648. doi: 10.1016/j.cemconcomp.2007.11.006

Kumagai S, Kurumatani N (2009) Asbestos fiber concentration in the area surrounding a former asbestos cement plant and excess mesothelioma deaths in residents. Am J Industr Med 52:790–798. http://onlinelibrary.wiley.com/doi/10.1002/ajim.20743/pdf

Li Z, Wang L, Wang X (2004) Compressive and flexural properties of hemp fiber reinforced concrete. Fibers Polymers 5:187–197. doi:10.1007/BF02902998

Li Z, Wang X, Wang L (2006) Properties of hemp fibre reinforced concrete composites. Compos A 37:497–505. doi:10.1016/j.compositesa.2005.01.032

Lima H, Willrich F, Barbosa N, Rosa M, Cunha B (2008) Durability analysis of bamboo as concrete reinforcement. Mater Struct 41:981–989. doi:10.1617/s11527-007-9299-9

Mesquita L, Czarnieski C, Filho A, Willrich F, Júnior H, Barbosa N (2006) Adhesion strength between bamboo and concrete. R Bras Eng Agr Ambient 10:505–516

Mohr B, Biernacki J, Kurtis K (2007) Supplementary cementitious materials for mitigating degradation of kraft pulp fiber cement-composites. Cem Concr Res 37:1531–1543. doi: 10.1016/j.cemconres.2007.08.001

Motta L, John V, Agopyan V (2009) Thermo-mechanical treatment to improve properties of sisal fibres for composites. 5th International Materials Symposium MATERIALS 2009—14th meeting of SPM, Lisbon

Passuello A, Moriconi G, Shah S (2009) Cracking behavior of concrete with shrinkage reducing admixtures and PVA fibers. Cem Concr Compos 31:699–704. doi:10.1016/j.cemconcomp.2009.08.004

Pehanich J, Blankenhorn P, Silsbee M (2004) Wood fiber surface treatment level effects on selected mechanical properties of wood fiber–cement composites. Cem Concr Res 34:59–65. doi:10.1016/S0008-8846(03)00193-5

Pimentel L, Beraldo A, Savastano H (2006) Durability of cellulose–cement composites modified by polymer. Engenharia Agricola 26:344–353

Powers RF (1999) On the sustainable productivity of planted forests. New Forests 17:263–306. doi:10.1023/A:1006555219130

Ramakrishna G, Sundararajan T (2005a) Impact strength of a few natural fibre reinforced cement mortar slabs: a comparative study. Cem Concr Compos 27:547–553. doi:10.1016/j.cemconcomp.2004.09.006

Ramakrishna G, Sundararajan T (2005b) Studies on the durability of natural fibres and the effect of corroded fibres on the strength of mortar. Cem Concr Compos 27:575–582. doi:10.1016/j.cemconcomp.2004.09.008

Rametsteiner E, Simula M (2003) Forest certification—an instrument to promote sustainable forest management? J Environ Manag 67:87–98. doi:10.1016/S0301-4797(02)00191-3

Razak A, Ferdiansyah T (2005) Toughness characteristics of Arenga pinnata fibre concrete. J Nat Fib 2:89–103. doi:10.1300/J395v02n02_06

Reis J (2006) Fracture and flexural characterization of natural fiber-reinforced polymer concrete. Constr Build Mater 20:673–678. doi:10.1016/j.conbuildmat.2005.02.008

Roma L, Martello L, Savastano H (2008) Evaluation of mechanical, physical and thermal performance of cement-based tiles reinforced with vegetable fibers. Constr Build Mater 22:668–674. doi:10.1016/j.conbuildmat.2006.10.001

Sample V (2006) Sustainable forestry and biodiversity conservation toward a new consensus. J Sustainable Forestry 21:137–150

Savastano H, Agopyan V (1999) Transition zone studies of vegetable fibre–cement paste composites. Cem Concr Compos 21:49–57. doi:10.1016/S0958-9465(98)00038-9

Savastano H, Warden P, Coutts R (2000) Brazilian waste fibres as reinforcement for cement-based composites. Cem Concr Compos 22:379–384. doi:10.1016/S0958-9465(00)00034-2

Savastano H, Warden P, Coutts R (2001a) Performance of low-cost vegetable fibre–cement composites under weathering. CIB World Building Congress, Wellington

Savastano H, Warden P, Coutts R (2001b) Ground iron blast furnace slag as a matrix for cellulose–cement materials. Cem Concr Compos 23:389–397. doi:10.1016/S0958-9465(00)00083-4

Savastano H, Warden P, Coutts R (2003) Mechanically pulped sisal as reinforcement in cementitious matrices. Cem Concr Compos 25:311–319. doi:10.1016/S0958-9465(02)00055-0

Savastano H, Warden P, Coutts R (2005a) Microstruture and mechanical properties of waste fibre–cement composites. Constr Build Mater 27:583–592. doi:10.1016/j.cemconcomp.2004.09.009

Savastano H, Warden P, Coutts R (2005b) Potential of alternative fibre cements as building materials for developing areas. Cem Concr Compos 25:585–592. doi:10.1016/S0958-9465(02)00071-9

Savastano H, Santos S, Radonjic M, Soboyejo W (2009) Fracture and fatigue of natural fiber-reinforced cementitious composites. Cem Concr Compos 31:232–243. doi:10.1016/j.cemconcomp.2009.02.006

Sedan D, Pagnoux C, Smith A, Chotard T (2008) Mechanical properties of hemp fibre reinforced cement: influence of the fibre–matriz interaction. J Eur Ceram 28:183–192. doi:10.1016/j.jeurceramsoc.2007.05.019

Silva J, Rodrigues D (2007) Compressive strength of low resistance concrete manufactured with sisal fiber. 51° Brazilian Congress of Ceramics. Salvador, Brazil

Silva F, Filho R, Filho J, Fairbairn E (2010) Physical and mechanical properties of durable sisal fiber–cement composites. Constr Build Mater 24:777–785. doi:10.1016/j.conbuildmat.2009.10.030

Stancato A, Burke A, Beraldo A (2005) Mechanism of a vegetable waste composite with polymer-modified cement (VWCPMC). Cem Concr Compos 27:599–603. doi:10.1016/j.cemconcomp.2004.09.011

Swamy R (1990) Vegetable fibre reinforced cement composites—a false dream or a potential reality? In Proc of the 2nd International Symposium on Vegetable Plants and their Fibres as Building Materials 3–8. Rilem Proceedings 7. Chapman and Hall

Swanson FJ, Franklin JF (1992) New forestry principles from ecosystem analysis of Pacific Northwest forests. Ecol Applic 262–274. www.whoi.edu/cms/files/jblythe/2005/6/forestecosystemanalysis_3587.pdf

Tonoli G, Joaquim A, Arsene M, Bilba K, Savastano H (2007) Performance and durability of cement based composites reinforced with refined sisal pulp. Mater Manufactur Process 22:149–156. doi:10.1080/10426910601062065

Tonoli G, Filho U, Savastano H, Bras J, Belgacem M, Lahr F (2009) Cellulose modified fibres in cement-based composites. Compos A 2046–2053. personales. http://www.upv.es/···/03···/tonoli%20composites%20part%20a.pdf

Tonoli G, Savastano H, Fuente E, Negro C, Blanco A, Lahr F (2010a) Eucalyptus pulp fibres as alternative reinforcement to engineered cement-based composites. Ind Crops Prod 31: 225–232. doi:10.1016/j.indcrop.2009.10.009

Tonoli G, Santos S, Joaquim A, Savastano H (2010b) Effect of accelerated carbonation on cementitious roofing tiles reinforced with lignocellulosic fibre. Constr Build Mater 24: 193–201. doi:10.1016/j.indcrop.2009.10.009

UNEP (2007) The last stand of the orangutan. State of emergency: Illegal logging, fire and palm oil in Indonesia's national Parks. In: Nellemann C, Miles L, Kaltenbom B, Virtue M, Ahlenius H (eds) United Nations Environment Programme, New York

Chapter 8
Earth Construction

8.1 General

There is no consensus about the date when man began to use earth for construction purposes. Minke (2006) mentioned that this may have happened over 9,000 years ago, grounding its beliefs on the fact that earth block (adobe)-based dwellings discovered in Turkmenistan dated from a period between 8000 and 6000 BC. Pollock (1999) refers that the use of earth for construction dates from the period of El-Obeid in Mesopotamia (5000–4000 BC). Berge (2009) mention that the oldest examples of adobe blocks, which were discovered in the Tigris river basin date back to 7,500 BC so earth construction could have been used for more than 10,000 years. It is not very relevant, whether the earth construction began more than 9,000 or over 10,000 years ago but it is not far from the truth that the earth construction began with early agricultural societies, a period whose current knowledge dates from 12000 BC to 7000 BC. There are countless cases of earth buildings built thousands of years ago that last to the twenty-first century. Even the Great Wall of China whose construction began about 3,000 years ago has extensive sections built on rammed earth. Evidences also show the use of earth construction by the Phoenicians in the Mediterranean basin including Carthage in 814 BC. The Horyuji Temple in Japan has rammed earth walls built 1,300 years ago (Jaquim 2008). This author refers to the existence of rammed earth-based buildings in the Himalayan region built in the twelfth century. Adobe-based building structures are common in Central America. The ruins of the city of Chanchán in Peru are among the most ancient earth based constructions (Alexandra 2006). The village of Taos in New Mexico is another example of ancient earth constructions (1000–1500 AC). Another good example is the city of Shibam in Yemen with earth buildings up to 11 floors that were built hundred years ago (Helfritz 1937). Currently almost 50% of the world's population lives in earth-based dwellings (Fig. 8.1).

The majority of earth construction is located in less developed countries, however, this kind of construction can also be found in Germany, France or even

F. P. Torgal and S. Jalali, *Eco-efficient Construction and Building Materials*,
DOI: 10.1007/978-0-85729-892-8_8, © Springer-Verlag London Limited 2011

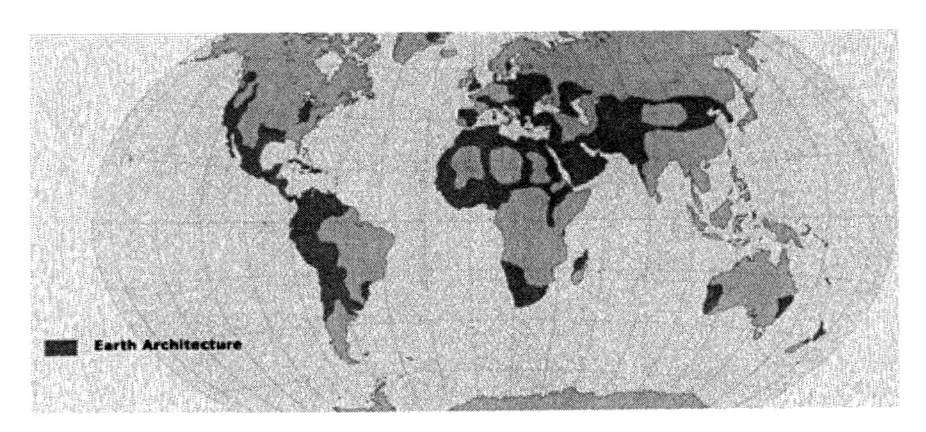

Fig. 8.1 Areas with earth construction

in the UK, a country that has an excess of 500,000 earth-based dwellings. MacDougall (2008) used interviews and site inspections to show that straw-bale construction and rammed-earth construction are gaining growing interest in the UK. The same author reveals that the lack of scientific data and the lack of experience by the mainstream construction industry in using these materials remain barriers to be overcome. Other authors (Desborough and Samant 2009) show that straw construction complies with building regulations and the UK climate, being a feasible option for this country. Williams et al. (2010) also show that thermal performance of earth block masonry meets current UK Building Regulation requirements. Earth construction has also increased substantially in the US, Brazil and Australia, largely due to the sustainable construction agenda, in which earth construction assumes a key role. In France investigations on earth construction were carried out by CRATerre, a laboratory founded in 1979 and linked to the School of Architecture in Grenoble, which acquired an institutional dimension in 1986 trough the recognition of the French Government. Houben et al. (2008) mention the success of an educational project undertaken in CRATerre, consisting of a scientific workshop with over 150 interactive experiences that in just 4 years had been attended by 11,000 visitors. As for Germany, Schroeder et al. (2008a) report the existence of vocational training in earth construction as well as courses that confer the expert title in this area. Three universities offer earth construction courses respectively, the University of Kassel, the University of Applied Sciences in Potsdam and the University of Weimar (Bauhaus). Earth construction is not only dependent on adequate training but also dependent on specific regulations. Several countries already have earth-construction-related standards. In Germany the first Earth Building Code dates back to 1944, but only in 1951 with DIN 18951, these regulations have been put into practice. In 1998 the German Foundation for the Environment disclosed several technical recommendations know as the "Lhmbau Regeln" (Schroeder et al., 2008a). Over the years they have been adopted by all the German states with the exception of Hamburg and Lower-Sáxony. A revised version of the "Lhmbau Regeln" was passed in 2008. Australia was one of the first countries to have specific regulations on earth construction. The Australian regulations were published in 1952 by the Commonwealth Scientific and Industrial Research Organization (CSIRO) under the designation of "Bulletin 5". This document has been revised in 1976, 1981, 1987 and 1992. In 2002 this document has been replaced by the Australian Earth Building Handbook (Maniatidis and Walker, 2003). In 1992 the Spanish Ministry of Transport and Public Works published a document entitled "Bases for design and construction with rammed earth" to support not only rammed earth but also adobe-based buildings (Maniatidis and Walker, 2003). Recently Delgado and Guerrero (2007) stated that earth construction is not yet regulated, posing several drawbacks such as the need to contract a building insurance during a 10 year warranty period. The United States has no specific regulations related to earth construction; but seismic regulations must be addressed by these constructions. Since 1991 New Mexico has a state regulation concerning rammed earth and adobe-based constructions (Maniatidis and Walker 2003). New Zeland has one of the most advanced legal regulations on earth construction which is structured in three distinct parts.

NZS 4297:1998—Engineering Design and Earth Buildings—Establishes performance criteria for mechanical strength, shrinkage, durability, thermal insulation and fire resistance;

NZS 4298:1998—Materials and Workmanship for Earth Buildings—Defines requirements for materials and workmanship;

NZS 4299:1998—Earth Buildings not Requiring Specific Design—This part is applicable for buildings with less than 600 m^2 (or 300 m^2 per floor) and provides constructive solutions for walls, foundations and lintels. In New Zealand the earth building regulations are dependent on the building height. For heights less than 3.3 m there is no need for a specific project, although the earth walls should respect the provisions of NZS 4298:1998. As for the buildings with a height between 3.3 and 6.5 m they shall be designed in accordance with NZS 4297:1998 (Jaquin 2008). Since 2001 Zimbabwe adopted a regulation based on the "Code of Practice for Rammed Earth Structures" (Keable 1996), which is composed by six sections: 1) Materials, 2) Formwork, 3) Foundations, 4) Wall design according to compressive strength, water absorption and erosion, 5) Masonry structural stability and 6) details and finishes. Shittu (2008) mentioned the following constraints of earth construction: lack of skilled craftsmanship; absence of earth-related courses and most of all the fact that earth construction is associated with a low-income status.

8.2 Techniques

Earth construction encompasses several techniques, the most usual being:

- Wattle and daub;
- Rammed earth (including earth projection);
- Earth bricks (adobe) or compressed earth blocks (CEB).

Fig. 8.2 Wattle and daub

In the wattle and daub technique the earth is pressed against a woven lattice of wooden strips (Fig. 8.2).

8.2.1 Rammed Earth

This technique means the compaction of moist earth (stabilized or not) inside a wooden formwork. In the earth projection technique the earth is previously stabilized and then it is projected against an inside formwork layer as happens in shotcrete works (Fig. 8.3).

The rammed earth technique requires a low amount of water, therefore it is suitable for regions where water is scarce. After placing the wooden formwork (Fig. 8.4) it is filled with a 10 cm earth layer followed by a ramming phase and then a new 10 cm layer are added and rammed. Afterwards, the wooden formwork is removed and then placed at a higher level until the desired hall height is reached.

Traditional compaction methods are made manually using heavy wooden tampers (Fig. 8.5a). This process requires quick tampering in order that the compression is performed with moist soil to obtain the desired cohesion. Modern rammed earth uses metal formwork and pneumatic rammers (Fig. 8.5b). Therefore, the time for compaction is shorter than in the traditional processes and is also less tiring. Middleton (1992) suggests the use of pneumatic rammers with circular heads and a diameter between 70 and 150 mm. The foundations of rammed earth walls are made of stone masonry or even concrete to prevent the rise of moisture by capillary action.

Regarding the minimum thicknesses of these walls, different authors suggest different values. According to Schroeder et al. (2008b) the German Earth Code mentioned a minimum of 32.5 cm for rammed earth walls.

Fig. 8.3 Earth projection

Fig. 8.4 Conventional wooden formwork

8.2.2 Adobe

Adobe is a very simple earth building technique being the most ancient . The word adobe comes from the Arab "attob" which means sun-dried brick (Rogers and Smalley 1995). The production of adobe bricks consists of filling wooden molds with moist earth which are then placed in the sun to dry. When the adobe dries shrinkage cracks could appear in its surface, so some authors (Neumann et al. 1984, Quagliarini and Lenci 2010) suggest the use of straw or other vegetable fibres to prevent this. However, this position is not unanimous because vegetable

Fig. 8.5 Rammers for earth construction: **a** Manual, **b** Pneumatic

Fig. 8.6 Adobe wall construction

fibres could rot leading to the appearance of fungi. Adobe masonry is very simple to build as it is conventional masonry (Fig. 8.6).

Adobe bricks are usually laid using an earth-based mortar to ensure greater adhesion between the different materials and prevent shrinkage cracks. Afterwards, the adobe masonry can be left plastered with an earth-based mortar or it can be without any covering.

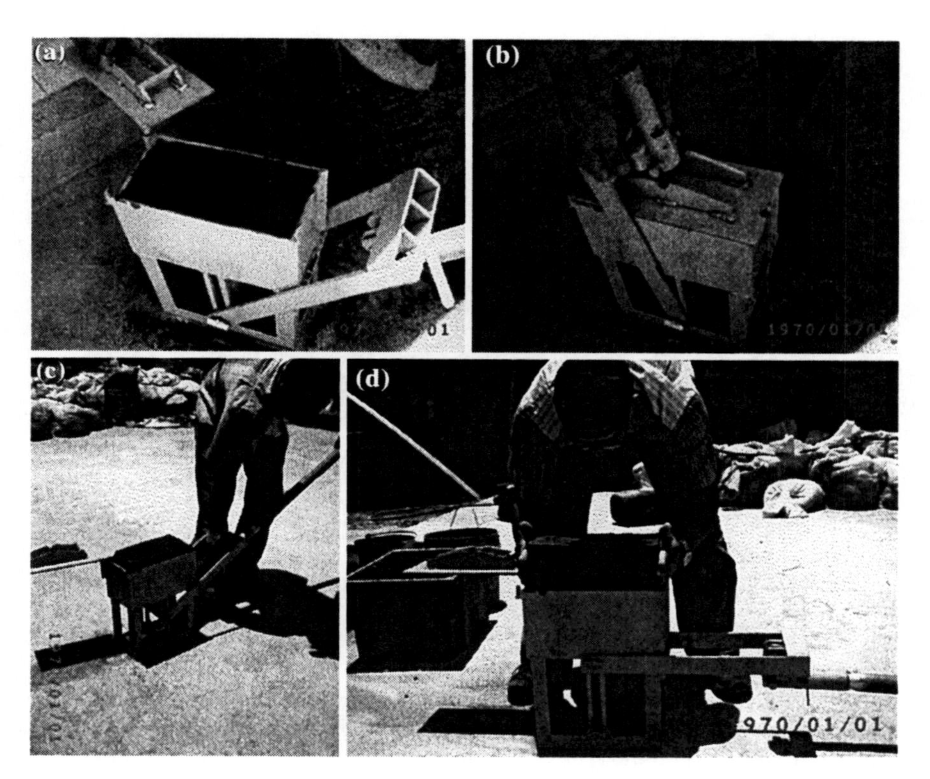

Fig. 8.7 Execution of a CEB with the Cinva-ram: **a** Filling the mold, **b** Compression of the earth, **c** Elevation of the block, **d** Removing the block

8.2.3 Compressed Earth Blocks

CEB represents an evolution of adobe bricks by using a specific device to compress the earth inside a mold. The pressure can be carried out manually or mechanically. The earth consistency is similar to that used in rammed earth allowing for the production of earth blocks that are heavier and more resistant than adobe bricks. The first machine used to make CEB was the CINVA-Ram created by Raul Ramirez in the International American Housing Centre (CINVA) in 1956 (Mukerji 1986). Figure 8.7 shows how to make a CEB with the CINVA-Ram.

Several other block making machines are also used like the Astram machine, developed in the mid 1970s at the Centre for Application of Science and Technology for Rural Areas in India, the CETA-Ram which is a modified CINVA-Ram developed in 1976 at the Centre of Appropriate Technical Experimentation in Guatemala (Mukerji 1986), the multi-block Brepak developed in 1980 at Building Research Establishment at Watford, England (Webb 1983), the CTA Triple-Block Press developed in 1982 at the Centre for Appropriate Technology in Paraguay and many others (Mukerji 1986). CEB produced manually require more manpower and

Fig. 8.8 Hydraulic machine for CEB production

are more time-consuming, however, they are cost-effective and could be made on-site reducing transportation costs. According to Shittu (2008) the success of the Cinva-Ram depends on the type of soil and sometimes the compression just expels the water leading to the disintegration of the earth block. When using a hydraulic machine (Fig. 8.8) the process is faster especially on machines that compress several blocks at once. These blocks show a higher mechanical resistance than the blocks made with Cinva-Ram and they also show a high resistance to water exposure. The mobility of hydraulic machines also allows for on-site production reducing transportation costs.

8.3 Earth Stabilization

The soil used in earth construction consists only in its mineral phase excluding the organic phase usually present in the first layers. This phase consists of mineral particles including clays, silts and sandy material, which are mixed together in varying proportions. The behavior of a soil depends not only on the amount of clay, silt and sand but also on the amount of water present in soil. This water encompasses the free water beneath the freatic level, the capillary water which is retained in the vicinity of the contact points of solids and finally the adsorbed water, i.e. the water held on the surface of the particles (with less than 0.002 mm) by electrochemical forces (Correia 1995). In order to understand the properties of the soil one must first characterize it using specific tests. The characterization of the mineral phase of a soil is carried out through tests that allow quantifying different types of properties, including its dimensions, its mechanical behavior and its deformation behavior for a certain level of humidity. These tests can be subdivided into two major groups, expedite tests to be held on-site that have a low level of confidence and laboratory tests made according to standard procedures.

8.3.1 On-Site Tests

These types of tests allow some initial conclusions about the type of soil available on-site, without the need for laboratory testing which is always expensive. The following tests are an adaptation of the tests used by the French group CRATerre and cited by Eusébio (2001):

a. Color observation: Soils with a high content of organic matter have a dark color. Pale soils mean the presence of feldspar or quartz sand. Red soils may be due to the presence of iron oxides.
b. Smell test: Soils with a high content of organic matter have a strong smell of humus, which is enhanced by heating or moistening of the soil.
c. Touch test: When rubbing a sample of soil between the hands, one perceives the presence of a sandy soil if it is rough. Plastic soils (when moist) indicate a high amount of clays.
d. Brigthness test: A ball of soil slightly moist and cut by a knife, will have opaque surface if there is a predominance of silt or will have a shiny surface if there is a predominance of clay.
e. Adherence test: In this test a ball of soil slightly moist is penetrated by a spatula. If the penetration is difficult and the earth sticks to the spatula, it is a clay soil. If the spatula enters and gets out easily is a sandy soil.
f. Sedimentation test: This test is an easy way to measure the sand, silt and clay content in a soil sample. First a bottle with 1 l of volume is filled up to ¼ of its capacity with soil and then it is filled with water. The bottle is shaken and allowed to stand for an hour, the procedure is repeated twice. After this procedure it is possible to measure the thickness of the layers of sand, silt and clay with a ruler.
g. Expedite sieving followed by visual test: This test uses loose dry soil and two sieves series ASTM, No. 200 (0.074 mm) and No. 10 (2 mm). The soil sample is passed on sieve No. 200 and the retained portion is passed on sieve No. 10. Comparing the volume of the soil that passed in each sieve it is possible to give a rough characterization: If the volume of the soil that passes sieve No. 200 is larger than the retained soil we are in a presence of a clayey soil. If the inverse situation occurs the soil is sand based. When using sieve No. 10, if the volume of the soil passed is lesser than the volume of the soil retained, the soil is a coarser one. When the inverse occurs the soil is sandy based.

h. Water retention test: A sample of soil is passed trough a sieve with a mesh of 1 mm. With a little volume of water a soil ball with a size of an egg is made. The soil ball is held in one hand and struck repeatedly with the other hand until the water appears on its surface. In a sandy soil ball the water appears on the surface after 5–10 strikes. A clayey soil or a silty clay requires 20–39 strikes. When there is no reaction the soil has a high clay content.

8.3.2 Laboratory Tests

Water content: In this test the mass of a sample of soil is compared before and after being dried in an oven at 105°C.

Organic matter: In order to estimate the organic matter content in the soil it must be heated at 400°C. Then one has to compare the mass differences before and after heating.

Particle size analysis: The test identifies the mass percentages of sandy soil above 0.074 mm (ASTM sieve No. 200), obtained by sieving it through a series of standard sieves. To identify the different constituents of the soil below 0.074 mm (silt and clay) it is necessary to use the sedimentation test. The soil is placed in a liquid suspension to determine the settling velocity which is dependent on the diameter of the particles through the Stokes law.

Atterberg limits: These limits allow knowing the behavior of a soil fraction below 0.4 mm according to their water content. They include the SL, the liquid limit (LL), the plastic limit (PL). The plasticity index (PI) is obtained by the difference PI = LL−PL. The shrinkage limit (SL) is the water content where further loss of moisture will not result in any more volume reduction. The LL is the water content obtained on the Casagrande device, a metal cup with a clay sample. In this test a groove was cut through the clay sample with a spatula, and the cup is repeatedly dropped 10 mm onto a hard rubber base during which the groove closes up gradually as a result of the impact. The moisture content at which it takes 25 drops of the cup to cause the groove to close over a distance of 13.5 mm (0.53 in) is defined as the liquid limit (LL). The PL is the water content that leads an earth cylinder with 3 mm diameter to crumble. If the cylinder crumbles with less than 3 mm it has too much water.

Proctor compactation test: The Proctor test is used to determine the water content that leads to the maximum density. This test uses a soil sample with less than 4.76 mm (Sieve No. 4) to which an increasing water content is added. The soil sample is compressed into three layers with 25 blows per layer, with a manual or mechanical device (with a mass of 2.49 kg falling by 30.5 cm). The soil density and the water content are registered allowing the determination of the minimum water content that leads to the maximum density. However, some authors argue that the Proctor test has a low compaction energy leading to a higher optimum water content than that recommended for rammed earth with a pneumatic

Table 8.1 Dry density after the Proctor compaction test (Doat et al. 1979)

Dry density (kg/m^3)	Soil classification
1,650–1,760	Mediocre
1,760–2,100	Very satisfactory
2,100–2,200	Excellent
2,200–2,400	Exceptional

Table 8.2 Atterberg limits of the soil used for earth construction (Doat et al. 1979)

	Recommended	Maximum and minimum
PI	7–18	7–29
LL	30–35	25–50
PL	12–22	10–25
SL	Optimum water content	8–18

Table 8.3 Soil plasticity classification (Doat et al. 1979)

Plasticity	PI
Week	5–10
Medium	10–20
High	>20

tamper. Maniatidis and Walker (2003) refer to an expedite way to obtain the optimum water content using a "drop test". A soil ball with a certain water content is dropped from a height of 1.5 m. If the ball does not break, the water content is excessive, if it breaks into several pieces then the water content is low.

Compressive strength test: This test is similar to the test used to assess the compressive strength of concrete or bricks. An earth specimen is submitted to an axial load until the rupture occurs.

8.3.3 Properties and Soil Classification

The CRATerre group classifies the soil according to its dry density after the Proctor compaction test (Table 8.1).

The same authors make some recommendations concerning the Atterberg limits of the soil used for earth construction (Table 8.2).

According to Michel (1976), the most suitable soils to be stabilised present low PI values. Doat et al. (1979) present a classification for the PI of a soil (Table 8.3).

The activity (A) of a soil or Skempton index is the PI divided by the percentage of clay-sized particles (lesser than 2 μm). A higher (A) means a higher deformability (Table 8.4).

Table 8.5 shows the characteristics of two soils used for earth construction.

Bahar et al. (2004) recommend that the optimum water content should be between 9.5% and 11%. According to the standard NZS 4298 the water content to be used in rammed earth should be located between 3% below the optimum water

Table 8.4 Clay activity (Doat et al. 1979)

Clay type	A = PI/(% clay < 0.002 mm)
Low activity	Ac < 0.75
Medium activity	0.75 < Ac < 1.25
Active	1.25 < Ac < 2.0
High activity	Ac > 2.0

Table 8.5 Characteristics of two soils used for earth construction

	Bahar (2004)	Guettala et al. (2006)
Clay and silt (%)	62	36
Sand (%)	38	64
LL (%)	39	31
PI	15	14
Optimum water content (%)	11	11.8
Dry density (kg/m^3)	1,760	1,877

content and 5% above (Hall and Djerbib 2004). The soils more suitable for earth construction should have a sand content between 50% and 70%. For adobe bricks Doat et al. (1979) recommend the following composition: sand (55% to 75%); silt (10% to 28%); clay (15 to 18%) and less than 3% of organic matter. Delgado and Guerrero (2007) mentioned that soils with a clay content between 10% to 20% should be used in CEB, and 10% to 15% in rammed earth. These authors also mentioned that independent of the technique used (rammed earth, adobe or CEB) the soils must have a minimum of 5% clay content. Jayasinghe and Kamaladasa (2007) mentioned high reductions in the compressive strength of lateritic soils stabilized with cement, when the fines percentage (clay and silt) exceeds 40%. Perera and Jayasinghe (2003) suggest that this percentage should not exceed 30%. Burroughs (2008) analyzed 104 soil types, compacted and stabilized with lime or cement in a total of 219 mixtures. According to this author a soil could be considered suitable for stabilization if its compressive strength exceeds 2 MPa (Fig. 8.9).

8.3.4 Particle Size Correction

When the soil does not exhibit the more favorable characteristics for earth construction it must be mixed with other soils in order to obtain the required characteristics.

If the soil is too clayey and very plastic, it must be mixed with sand or with a sandy-like soil, however, if it is very sandy it must be corrected through the addition of fine particles. If the soil contains a high amount of coarse particles, it must it be passed by a sieve with the proper mesh. If the soil has too many fine particles, the solution may undergo washing, however, this operation may remove all the fines, so it is preferable to mix it with a sandy soil.

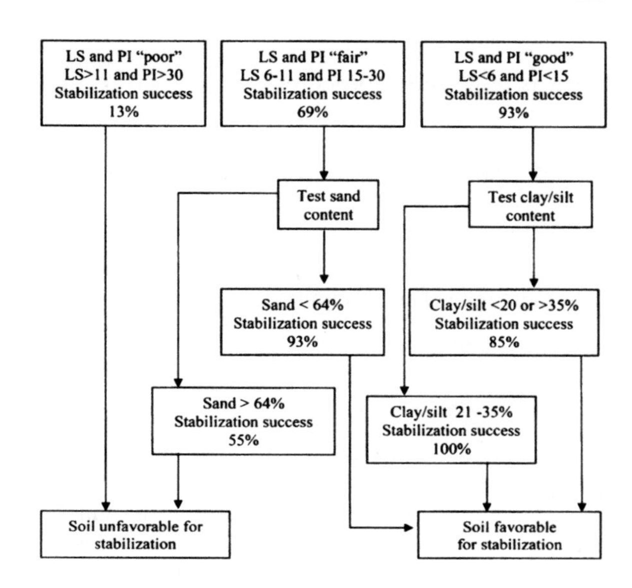

Fig. 8.9 Procedures for determining soil favorability for stabilisation (Burroughs 2008)

8.3.5 Soil Stabilization

The soil stabilization means changing the soil characteristics in order to improve its mechanical or physical behavior. The stabilization process aims at the reduction of the soil plasticity, the improvement on its workability and the deflocculation resistance and also the resistance to erosion. The methods for soil stabilizing can be subdivided into:

- Mechanical stabilization which seeks to improve the characteristics of the soils through a higher density.
- Chemical stabilization with lime, cement or other additions.

Anger et al. (2008) studied the cohesion mechanism of the earth. According to these authors, water is one of the main factors responsible for earth cohesion, due to its surface tension. As to the clay matrix it consists of lamellar microscopic particles whose cohesion is due to the nanoscale capillary connections. The optimum stabilization process encompasses two different stages: the first related to the dispersion of clay induced by electrostatic repulsion that minimizes the water content and reduces the final porosity, and the second, consisting of a binding mechanism. These authors report the existence of various cements available in nature, such as "silcrete" resulting from the dissolution and hardening of silica, "ferricrete" resulting from the agglomeration of sand and other aggregates through the action of iron oxide due to the oxidation of percolating solutions containing iron salts. The stabilization of soils for earth construction can include vegetable fibres (Ghavami et al. 1999), artificial fibres (Binici et al. 2005) or even animal droppings (Ngowi 1997). More recently Silva et al. (2010) studied the nests of andorinha-dos-beirais birds concluding that a mixture of clay and polysaccharide/sugar is responsible for its high strength and high durability.

8.3.5.1 Stabilization with Lime

Mixing lime into a moist soil generates various chemical reactions which cause the agglutination of the soil particles and the modification of their characteristics. The most important reactions during the lime stabilization process are as follows:

- Ion exchange and flocculation;
- Cementing action (or pozzolanic reaction);
- Carbonation.

The ion exchange causes the Ca^{2+} cations to adsorb into the surface of the particles decreasing their electronegativity and promoting flocculation. The action of calcium ions begins immediately after the addition of lime to the soil, leaving a loose moist mixture in a curing process (a process also named as rotting due to its smell), a decrease in the plasticity of the soil occurs, which then becomes brittle and easily breaks up. To achieve these benefits all that is needed is a small amount of lime. The cementing action requires considerable time and is therefore a slow reaction, being responsible for long-term action of lime stabilization. It is designated as pozzolanic reaction and occurs under hot weather and can be accelerated by using suitable additives. Through the reaction between lime, silica and aluminum present in the clay particles promote the formation of calcium silicate hydrates and/or calcium aluminates. The interaction between lime and clay dissolves the silica and aluminum particles under the high pH created by the molecules of $Ca(OH)_2$. The dissolved materials combine with calcium ions, forming cementitious products that connect the clay particles. Finally, the carbonation reaction is the reaction of lime with the carbon dioxide from the atmosphere. It consists in the chemical alteration of clay minerals due to the reaction of atmospheric carbonate ions with calcium ions to form calcium carbonate. This is the reverse reaction that occurs in the production of lime from limestone, and should be avoided since the formed calcium and magnesium carbonates affect the pozzolanic reaction, preventing the required soil strength. The identification of physical and chemical properties of lime is essential for its use in soil stabilization. One of the main properties to consider is its particle size, since it affects various properties of the soil–lime mixture, such as the hydration speed, the density and also the homogeneity. The particle size of lime is conditioned by the particle size of the limestone rock, by the calcination process, by the final product (slaked lime) and even by possible additional mill operations. The knowledge of the lime fineness may be useful for the evaluation of the degree of homogenization and the reaction between lime, soil and water, because larger areas of contact give rise to more balanced mixtures. The porous structure of the lime particles causes its outer surface to be in contact with water. Through absorption and adsorption phenomena, a part of its interior surface is also surrounded by water. The high reactivity shows the high speed of the lime action after being mixed with the soil and the efficiency of its stabilizer action. This property allows to anticipate the duration of the reactions and if they are exothermic, the increase in the temperature. The production of soil–lime mixtures for soil stabilization needs appropriate lime content. The optimum lime content depends on the future application of the stabilized soil. These may aim at the decrease of the plasticity and workability increase—improvements—making permanent changes that alter the strength of the mixture—stabilization. For the composition of soil–lime mixtures the Atterberg limits are determined, the particle size and the soil classification. Afterwards, compaction tests and resistance and durability tests take place. In cold climates, durability constitutes a major requirement. The content of lime to be used in soil stabilizing, should be the in the range of 1 to 10%, however, the exact percentage must be determined for each case. The use of higher amounts would not be economical nor necessary, but one should never use less than 3%, because even if in the laboratory the mixture with a lower percentage has achieved the desired properties, one must remember that the mixing conditions on-site are somehow different. The lime stabilization process is suitable for soils with a fine fraction very plastic and very expansive. The material initiates the cementing process, strengthens and becomes more granular and then it can be regarded as a material with particles of larger size and greater friction angle. Millogo et al. (2008) studied the effect of adding lime to clay soils for the manufacture of adobe bricks, mentioning that the use of a lime content of 10% maximizes the compressive strength and minimizes the water absorption of the bricks. According to these authors, the addition of increasing percentages of lime induces the formation of calcite and CSH phases generated by the reaction between lime and quartz (silica) of the soil. But when the lime percentage rises to 12% the formation of portlandite also occurs.

8.3.5.2 Soil Stabilization with Cement

Stabilization with cement involves two different mechanisms on the cement content that is added to the soil (Cruz and Jalali 2009). When low cement contents are used a modification of the clay fraction of the soil takes place. This phenomenon decreases its plasticity, and sometimes it can lead to an increase in the mechanical strength, because the hydration of cement particles will contribute to form independent nuclei. During the hydration process some calcium hydroxide will be formed reacting with the clay particles to form CSH, however, its volume will be a low one. The hydration of cement particles will increase with time leading to increased soil strength. The particle size and plasticity leads to different stabilization mechanisms (cementing action or modifying action) that depends on the cement content being simultaneous or not. In soils with a relevant sand fraction the cement will not be enough to fill all the empty spaces. For these soils the cement will link the contact areas of the soil particles. Since these areas depend on the soil particle size, maximizing the number of contact points, lowers the cement

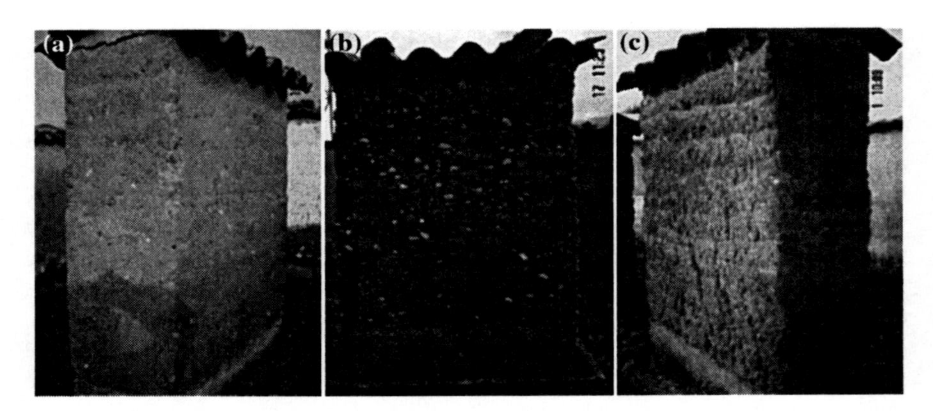

Fig. 8.10 Rammed earth masonry sections exposed during 20 years to natural climatic conditions. **a** Wall made with soil stabilized with 5% lime, **b** Wall made with soil without stabilization (mixed soil), **c** Wall made with soil without stabilization (Bui et al. 2008)

content required to reach a certain strength level. It is worth remembering that soil stabilization with cement is strongly affected by the presence of organic matter in the soil, this slows or inhibits the cementing action preventing the release of calcium ions. As for the volume of water necessary for soil–cement mixtures it matches the optimum water content for maximum compaction, obtained in the Proctor test. The water needed for cement hydration is less than the amount needed for maximum compaction, so that the water necessary for the hydration process is ensured as long as no loss occurs during the curing period. The water content of soil–cement mixtures, should be in the range of 0.95 to 1.10 times the water content for maximum compaction (Pereira 1970).

8.4 Durability

What is known in terms of the durability of earth construction comes from the fact that some of these buildings last for hundreds of years. Durability has also been assessed by accelerated aging tests and more recently from monitoring experimental sections of earth masonry sections built a dozen years ago. The main mechanism responsible for the erosion of earth walls have to do with the kinetic energy of the impact of rainfall (Heathcote 1995). This justifies the worst durability behavior of earth walls oriented to the South, a direction usually associated with wind-based rain. Other authors (Ogunye and Boussabaine 2002) mentioned that rain does not always have a erosive effect on the earth walls which only happens for rain intensities above 25 mm/m. Bui et al. (2008) evaluated the performance of 104 sections of rammed earth masonry sections with and without stabilization, which were exposed for 20 years to natural climatic conditions (Fig. 8.10).

The durability of earth buildings is also dependent on appropriate maintenance and repairs that are compatible with the original construction (Little and Morton,

Table 8.6 Assessment of the durability of earth constructions (Heathcote 2002)

Test	Types of tests		
	Indirect	Simulation	Accelerated erosion
Compressive strength	x		
Superficial strength	x		
Permeability	x		
Erosion	x		
Drip tests	x	x	
Spray tests			x

Table 8.7 Accelerated erosion spray tests (Maniatidis and Walker 2003)

	Distance (mm)	Pressure (kPa)	Nozzle	Time (min)
Israel (Cytryn, 1955)	250 vertical	50	Spray	33
Austrália-CSIRO	470 vertical	50	Spray	60
Dep. Housing Washington	175 horizontal	137	Shower	120
Norton	180 horizontal	137	Shower	120
Houben and Guillaud	200 horizontal	140	Shower	120

2001). The assessment of the durability of earth constructions can be made indirectly through the analysis of the compressive strength and permeability of earth specimens. It can also be made using erosion tests by mechanical impact or water falling in a drop by drop mode (Table 8.6).

Geelong test—this test was specially designed for adobe specimens (Walker 2000). The test consists to drip water at a controlled rate onto earth bricks (inclined 30°) from a height of 400 mm. The test ends when the volume of dripped water reach 100 ml, this should happen after 30 min. The degree of erosion is given by the depth of the erosion caused by the drop of water. A depth greater than 15 mm means that the earth bricks must be rejected.

Accelerated erosion test (SAET)—this test also uses a similar inclined 30° earth specimen but uses a jet of water rather than individual drops used in the Geelong test. The SAET results are obtained from the pitting depth caused by the falling water, and the specimens with a depth of over 30 mm are considered unsuitable for earth construction.

In the last 50 years several accelerated erosion tests based on spraying water horizontally onto earth specimens were developed (Table 8.7).

Bulletin 5 accelerated erosion test—this test was developed in Australia in the early 1980s and was named after the document in which it is contained. This test consists of spraying water horizontally onto an earth specimen during 1 h or until the water penetrates the earth specimen. The test uses a water pressure of 50 kPa which corresponds to a velocity of 10 m/s. After 15 min the test is interrupted for the assessment of the erosion rate. The final erosion assessment is given in millimeter per minute and the maximum acceptable erosion assessment is 1 mm/min.

Fig. 8.11 Bulletin 5 accelerated erosion test: **a** Spray, **b** Specimen after Bulletin 5 test (Heathcote and Moore 2003)

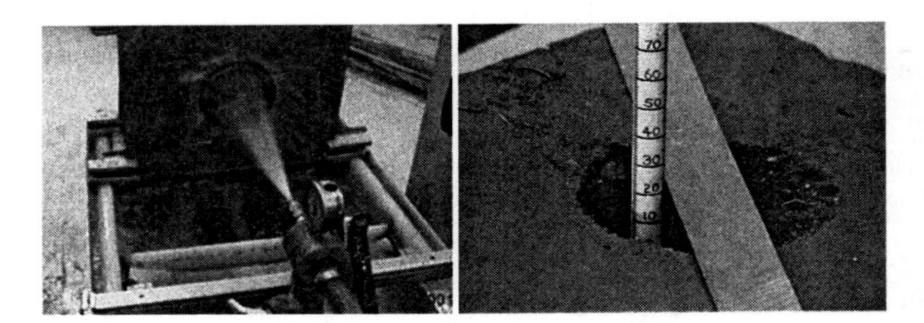

Fig. 8.12 UTS spray test: Left spray, right erosion assessment (Heathcote and Moore 2003)

According to Heathcote and Moore (2003) this test can hardly reproduce the action of rainfall because it has a very intense effect leaving bore holes in the earth specimens (Fig. 8.11b).

University of Technology Sydney (UTS)—the UTS spray test is a refinement of the Bulletin 5 test, it uses a higher water pressure (70 kPa) producing an even more erosive effect, that is less concentrated due to the use of a specific kind of nozzle capable of producing a turbulent flow with a velocity similar to the wind velocity (9 m/s) that occur in Sydney on rainy days (Fig. 8.12).

8.5 Eco-Efficiency Aspects

8.5.1 Economic Advantages

For less-developed countries the cost-efficiency aspect remains of paramount importance. Zami and Lee (2010) quote several authors for whom "earth construction is economically beneficial", nevertheless one cannot take this as a guaranteed truth because the economics of earth construction depend on several aspects such as: construction technique, labour costs, stabilization process, durability and repair needs. Williams et al. (2010) mentioned that the materials used in earth construction in UK have no significant impact in the final cost. These authors state that the production and construction costs represent the most important part because earth construction is labor intensive. However, this is not the case in less-developed countries in which labor is available for a very low cost. According to Sanya (2007) this is very important to create decentralized job creation. In these countries the cost-efficiency is dependent on the nature and the amount of the binder used in the stabilization process. Recent investigations (Zami and Lee 2010) show that soil stabilization with gypsum shows to be much more cost-effective than with Portland cement.

8.5.2 Non-Renewable Resource Consumption and Waste Generation

The use of soil for earth construction cannot be regarded as the use of a renewable resource; however, one must recognize that it is very different from the extraction of raw materials needed for the construction materials used in conventional masonry. This is because generally the soil used in earth construction is located immediately below the organic layer of the soil. Therefore, earth extraction generally involves the removal of the top layer of the soil, an operation without high energy needs since it can be done manually. If we assume that the building is made with soil located in its vicinity there is no pollution associated with its transportation. This is very different from conventional masonry in which concrete blocks and fired-clay bricks are always very distant from construction sites thus implying high transport distances responsible for the emission of GHGs. Regarding earth construction wastes they can simply be deposited at the site of its extraction without any environmental hazard involved. Even when the soil is stabilized with cement or lime, it can be reused in this type of construction, so we may consider that earth construction hardly generates any waste. As a comparison the traditional fired-clay brick masonry implies a relevant amount of waste because the use of broken pieces takes place quite often in this kind of masonry. According to Morton (2008) earth construction could reuse the 24 million tones of waste soil produced every year in UK.

8.5.3 Energy Consumption and Carbon Dioxide Emissions

Some authors compared the carbon dioxide emissions of earth blocks and the emissions of the construction materials used in conventional masonry, showing the good environmental performance of the former (Fig. 8.13).

For a house with three rooms and an area of 92 m^2 made with earth walls the values in Fig. 8.13 represent a reduction of 7 tons of CO_2 compared to fired clay

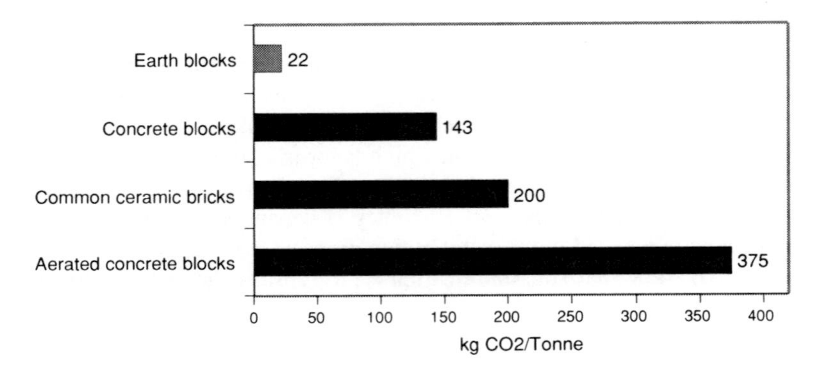

Fig. 8.13 Embodied carbon in different masonry materials (Morton et al. 2005)

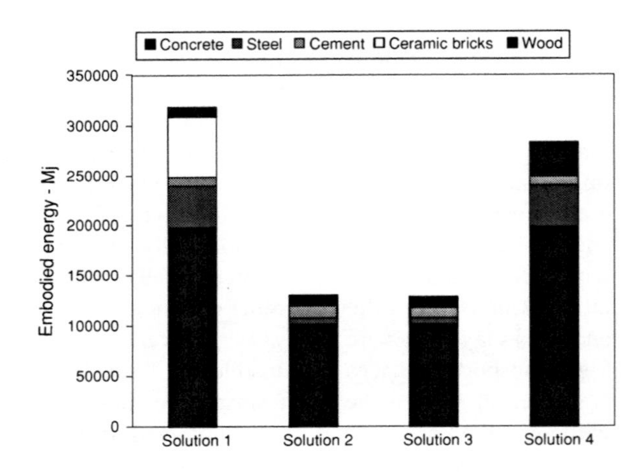

Fig. 8.14 Embodied energy for four different scenarios (Lourenço 2002)

brick and a reduction of 14 tons of CO_2 if aerated concrete blocks were used. Lourenço (2002) studied the embodied energy (wood, concrete, steel, fired-clay bricks and cement) of a single floor building comprising the following variants:

- Solution 1: Building with a reinforced concrete structure, fired-clay bricks masonry and roof slab using precast reinforced concrete beams and fired-clay hollow elements;
- Solution 2: Building with CEB masonry with top concrete beams and wooden roof;
- Solution 3: Building with exterior walls made on rammed earth, interior walls made on adobe and wooden roof;
- Solution 4: Building with a reinforced concrete structure and adobe walls.

This author shows that the embodied energy of earth buildings (Solutions 2 and 3) is half the embodied energy of a conventional construction (Fig. 8.14).

According to Morton (2008) the replacement of only 5% of concrete blocks used in the UK masonries by earth masonry would mean a reduction in CO_2 emissions of approximately 100,000 tons. Shukla et al. (2008) studied adobe-based buildings observing an embodied energy of 4.75 GJ/m^2. Nevertheless, compressed-stabilized earth blocks are more eco-friendly than fired-clay and their manufacture consumes less energy (15 times less) and pollutes less than fired-clay bricks (8 times less) Zami and Lee (2010). Reddy and Kumar (2010) show that the embodied energy in cement-stabilized rammed earth walls increases linearly with the increase in cement content and is in the range of 0.4 to 0.5 GJ/m^3 for a cement content in the range of 6% to 8%. Lax (2010) assessed the LCA of rammed earth showing that the stabilization with cement makes the embodied carbon to raise from 26 Kg CO^2 to 70 Kg CO^2.

8.5.4 Indoor Air Quality

Earth construction is not associated with the adverse effects of indoor air volatile organic compounds (VOCs) so the occupants of these buildings will have a superior air quality (Wargocki et al. 1999). Another advantage of the indoor air quality of earth buildings relates to its ability to control the relative humidity (Minke 2000). Some investigations show that the earth blocks are capable of absorbing ten times more weight moisture than fired-clay bricks (Fig. 8.15). Earthen structures act as a relative humidity flywheel, equalizing the relative humidity of the external environment with that of the pores within the walls (Jaquin 2008; Allison and Hall 2010). According to Morton (2008) the hygroscopic behavior of construction materials can be more effective in reducing the indoor air relative humidity than the use of ventilation. This author mentioned a study conducted in Britain where it was noticed that earth construction is capable of keeping the relative humidity of indoor air between 40% and 60%, this range being the most suitable for human health purposes. Recently, Allison and Hall (2010) showed that earth walls have a high potential to stabilize the relative humidity of indoor air. High levels of humidity above 70% are responsible for the appearance of molds which can trigger allergic reactions (Arundel et al. 1986). Relative humidity values above 60% are associated with the presence of mites and also asthmatic diseases (Howieson 2005).

On the other hand, a relative humidity below 40% is linked to the syndrome of "sick buildings" typical of very dry indoor air. This leads to a drying of the respiratory mucosa, resulting in respiratory diseases such as tonsillitis, pharyngitis or bronchitis. Therefore, it is easily understood that public health statistics in recent decades show an increase of almost 50% in the occurrence of health problems from respiratory conditions such as asthma (Heerwagen 2000). Berge (2009) mentioned that the Hospital of Feldkirch in Austria in which a 180 m gallery was built with long sections coated with rammed earth (in some cases up to 6 m high), with the sole aim of achieving the stabilization of the relative humidity without using conventional mechanical devices (Fig. 8.16).

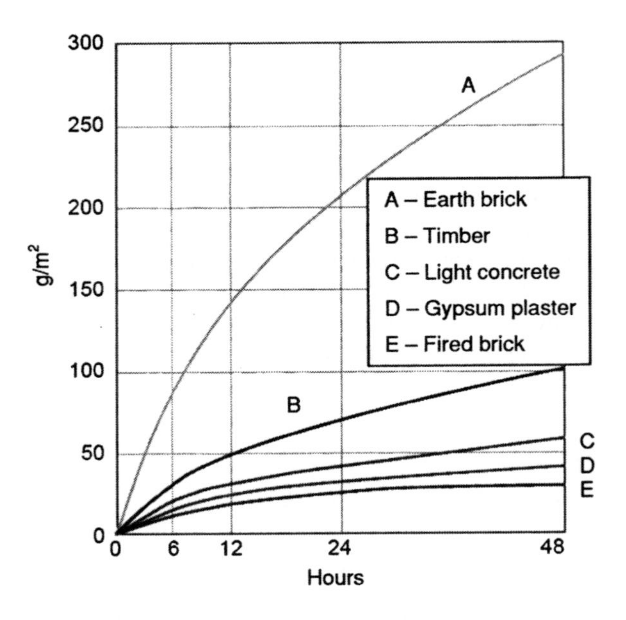

Fig. 8.15 Weight of moisture absorbed by different materials when relative humidity increases from 50% to 80% (Minke 2000)

Fig. 8.16 Rammed earth wall, Hospital of Feldkirch, Áustria (Berge 2009)

8.6 Conclusions

Earth construction exists since the early agricultural societies, a period whose current knowledge dates from 12000 to 7000 BC. There are countless cases of earth constructions built thousand years ago that made it to the twenty-first century.

Nowadays, the majority of earth construction is located in less-developed countries. Unfortunately, the fact that earth construction is associated with low-income status is probably one of the most important reasons explaining why less-developed countries try to emulate the use of polluting construction techniques based on reinforced concrete and fired-clay bricks. Earth construction can also be found in developed countries, where a growing awareness on the importance of this type of construction can be witness nowadays. Although this construction technique is cost-effective its economic advantages are dependent on the nature and the amount of the binder used in the stabilization process. Soil stabilization with gypsum shows to be much more cost-effective than that with Portland cement. Earth construction is associated with low-embodied energy, low carbon dioxide emissions and very low-pollution impacts. The use of cement for soil stabilization increases embodied energy. Therefore further studies about the environmental performance of earth construction stabilized with non-Portland cement binders are needed. The use of pozzolanic potential aluminosilicate wastes could also be analyzed. Earth construction is also responsible for an indoor air relative humidity beneficial to the human health; therefore, earth construction has clear competitive advantages in the field of eco-efficiency over conventional construction assuring it a promising future in the years to come.

References

Alexandra S (2006) Architecture and earth construction in Piaui: research, characterization and analysis. Master Thesis, Federal University of Piauí, Brasil

Allison D, Hall M (2010) Hygrothermal analysis of a stabilised rammed earth test building in the UK. Energy Build 42:845–852. doi:10.1016/j.enbuild.2009.12.005

Anger R, Fontaine L, Houben H, Doat P, Damme H, Olagnon C, Jorand Y (2008) Earth: a concrete like any other? 5th International Conference on Building with Earth-LEHM 2008, 59–65, Weimar, Germany

Arundel A, Sterling E, Biggin J, Sterling T (1986) Indirect health effects of relative humidity in indoor environments. Environ Health Perspect 65: 351–361. ehp.niehs.nih.gov/members/1986/065/65050.PDF

Bahar R, Benazzoug M, Kenai S (2004) Performance of compacted cement-stabilised soil. Cem Concr Compos 26:811–820.10.1016/j.cemconcomp.2004.01.003

Berge B (2009) The ecology of building materials, 2nd edn. Architectural Press, Oxford

Binici H, Aksogan O, Shah T (2005) Investigation of fibre reinforced mud brick as a building material. Constr Build Mater 19:313–318.10.1016/j.conbuildmat.2004.07.013

Bui QB, Morel JC, Venkatarama BV, Ghayad W (2008) Durability of rammed earth walls exposed for 20 years to natural weathering. Build Environ 44: 912–919.10.1016/j.buildenv.2008.07.001

Burroughs S (2008) Soil property criteria for rammed earth stabilization. J Mater Civil Eng 265–273. 10.1061/(ASCE)0899-1561(2008)20:3(264)

Correia AGC (1995) Soil mechanics and foundations I-Theoretical elements. IST-UTL, Lisbon

Cruz ML, Jalali S (2009) Soil stabilization with cement.TecMinho, Guimarães, Portugal

Delgado MC, Guerrero IC (2007) Earth building in Spain. Constr Build Mater 20:679–690. 10.1016/j.conbuildmat.2005.02.006

Desborough N, Samant S (2009) Is straw a viable building material for housing in the United Kingdom? Sustainability 2: 368–374. doi:10.1089/SUS.2009.9814

Doat P, Hays A, Houben H, Matuk S, Vitoux F (1979) Construire em terre. France CRAterre-École d'Architecture de Grenoble
Eusébio APJ (2001) Retroffiting and improvement of rammed earth walls. Master Thesis, UTL-IST, Lisbon, Portugal
Ghavami K, Filho R, Barbosa N (1999) Behaviour of composite soil reinforced with natural fibres. Cem Concr Compos 21:39–48. doi:10.1016/S0958-9465(98)00033-X
Guettala A, Abibsi A, Houari H (2006) Durability study of stabilized earth concrete under both laboratory and climatic conditions exposure. Constr Build Mater 20:119–127.10.1016/j.conbuildmat.2005.02.001
Hall M, Djerbib Y (2004) Moisture ingress in rammed earth: Part 1–the effect of soil particle-size distribution on the rate of capillary suction. Constr Build Mater 18:269–280. 10.1016/j.conbuildmat.2003.11.002
Heathcote KA (1995) Durability of earthwall buildings. Constr Build Mater 9:185–189. doi: 10.1016/0950-0618(95)00035-E
Heathcote KA (2002) An investigation into the erodibility of earth wall units. PhD Thesis, University of Technology, Sydney
Heathcote K, Moore G (2003) The UTS durability test for earth wall construction. www.dab.uts.edu.au/ebrf/research/leipzig-paper.doc
Heerwagen J (2000) Green buildings, organizational success and occupant productivity. Build Res Infor 28:353–367. doi:10.1080/096132100418500
Helfritz H (1937) Land without shade. J R Central Asian Soc 24:201–216
Houben H, Doat P, Fontaine L, Anger R, Aedo W, Olagnon C, Damme H (2008) Builders grains–a new Pedagogical tool for Earth architecture education. 5th international conference on building with Earth-LEHM 2008, 51–57, Weimar, Germany
Howieson S (2005) Housing and asthma. Spon Press, London
Jaquin PA (2008) Analysis of historic rammed earth construction. PhD Thesis. Durham University, UK
Jayasinghe C, Kamaladasa N (2007) Compressive strength of cement stabilized rammed earth walls. Constr Build Mater 21:1971–1976. doi:10.1016/j.conbuildmat.2006.05.049
Keable J (1996) Rammed earth structures. A code of practice. Intermeadiate Technology, London, UK
Lax C (2010) Life cycle assessment of rammed earth. Master Thesis, University of Bath, United Kingdom. http://www.ebuk.uk.com/docs/Dissertation_Clare_Lax.pdf (accessed in June 22, 2011)
Little B, Morton T (2001) Building with earth in Scotland: Innovative design and sustainability. Scottish Executive Central Research Unit, Edinburgh, Scotland
MacDougall C (2008) Natural building materials in mainstream construction: Lessons from the U.K. J Green Build 3:3–14. doi: 10.3992/jgb.3.3.1
Lourenço P (2002) Earth constructions. Master Thesis, UTL-IST, Lisbon
Maniatidis V, Walker P (2003) A Review of rammed earth construcion. University of Bath
Middleton GF (1992) Bulletin 5, Earth wall construction. CSIRO division of building, 4th edn. Construction and Engineering, North Ryde
Millogo Y, Hajjaji M, Ouedraogo R (2008) Microstructure and physical properties of lime-clayey adobe bricks. Constr Build Mater 22:2386–2392.10.1016/j.conbuildmat.2007.09.002
Minke G (2000) Earth construction handBook. The building material earth in the modern architecture. WIT Press, Southampton
Minke G (2006) Building with earth, design and technology of a sustainable architecture. Birkhäuser–Publishers for Architecture, Basel
Morton T (2008) Earth masonry–design and construction guidelines. HIS BRE Press
Morton T, Stevenson F, Taylor B, Smith C (2005) Low cost earth brick construction: monitoring and evaluation. Arc, Architects
Mukerji K (1986) Soil block presses. Publication of Deutsches Zentrum fr Entwicklungstechnologien-GATE, a Division of the Deutsche Gesellschaft fr Technische Zusammenarbeit (GTZ), GmbH

Neumann JV, Bernales JB, Blondet M (1984) Seismic resistance of adobe masonry. Pontificia University Católica of Peru

Ngowi A (1997) Improving the traditional earth construction: a case study in Botswana. Constr Build Mater 11:1–7.10.1016/S0950-0618(97)00006-8

Ogunye FO, Boussabaine H (2002) Diagnosis of assessment methods for weatherability of stabilized compressed soil blocks. Constr Build Mater 16:163–172. doi:10.1016/S0950-0618(02)00004-1

Pereira OA (1970) Soil stabilization. Course 108, LNEC, Lisbon

Perera A, Jayasinghe C (2003) Strength characteristics and structural design methods for compressed earth block walls. Masonry Inter 16:34–38

Pollock S (1999) Ancient Mesopotamia. Cambridge University Press, Cambridge

Quagliarini E, Lenci S, Iori M (2010) Mechanical properties of adobe walls in a Roman Republican domus at Suasa. J Cult Herit 11: 130–137. doi:10.1016/j.culher.2009.01.006

Reddy B, Kumar P (2010) Embodied energy in cement stabilised rammed earth walls. Energy and Buildings 42: 380–385. doi:10.1016/j.culher.2009.01.006

Rogers CDF, Smalley IJ (1995) The adobe reaction and the use of loess mud in construction. Constr Build Mater 40:137–138.10.1016/0013-7952(95)00064-X

Sanya T (2007) Living Earth. The sustainability of Earth Architecture in Uganda. PhD Thesis. The Oslo School of Architecture and Design, Norway

Silva B, Correia J, Nunes F, Tavares P, Varum H, Pinto J (2010) Bird nest construction-Lessons for building with earth. WSEAS Transactions on Environment and Development 6, 95–104. http://www.wseas.us/e-library/transactions/environment/2010/89-158.pdf

Schroeder H, Rohlen U, Jorchel S (2008a) Education and vocational training in building with earth in Germany. In 5th international conference on building with earth-LEHM 2008, 193–197, Weimar

Schroeder H, Volhard F, Rohlen U, Ziegert C (2008b) The "Lehmbau Regeln" in 2008—a review after 10 years of use in practice. In 5th international conference on building with earth-LEHM 2008,13–21, Weimar

Shittu T (2008) Earth building norms and regulation: a review of Nigerian building codes. In 5th international conference on building with Earth-LEHM 2008, 41–47, Weimar

Shukla A, Tiwari G, Sodha MS (2008) Embodied energy analysis of adobe house. Renew Energy 34:755–761.10.1016/j.renene.2008.04.002

Walker P (2000) Review and experimental comparison of erosion tests for earth blocks. In: Terra 2000, Proceeding of the 8th international Conference on the study and conservation of earthen architecture. Icomos, Torquay, pp 176–181

Wargocki P, Wyon D, Baik Y, Clausen G, Fanger P (1999) Perceived air quality, sick building syndrome (SBS) symptoms and productivity in an office with two different pollution loads. Indoor Air 9:165–179

Webb D (1983) Stabilized Soil Construction in Kenya. Proceedings, International Conference on Economical Housing in Developing Countries, UNESCO, Paris

Williams C, Goodhew S, Griffiths R, Watson L (2010) The feasibility of earth block masonry for building sustainable walling in the United Kingdom. J Build Apprais 6: 99–108. doi:10.1057/jba.2010.15

Zami M, Lee A (2010) Stabilised or un-stabilised earth construction for contemporary urban housing? 5th International Conference on Responsive Manufacturing 'Green Manufacturing'. Ningbo Higher Education Park, the University of Nottingham Ningbo, China, 11–13 January

Zami M, Lee A (2010) Economic benefits of contemporary earth construction in low-cost urban housing-State-of-the-art review. J Build Apprais 5:259–271. doi:10.1057/jba.2009.32

Chapter 9
Durability of Binder Materials

9.1 General

The importance of durability in the context of the eco-efficiency of construction and building materials has been rightly put by Mora (2007), when he stated that increasing the durability of concrete from 50 to 500 years would mean a reduction of its environmental impact by a factor of 10. Materials with low durability require frequent maintenance and conservation operations or even its integral replacement, being associated with the consumption of raw materials and energy. It is clear that a material with high embodied energy but with a high durability can be environmentally preferable to a material with a lower embodied energy but with a much lower durability. Similar considerations can be made about recyclable construction materials, because although it is positive to reuse materials and materials containing wastes it is also important to assess whether its low durability does not annul its environmental advantages.

9.2 Pathology and Durability

9.2.1 Concrete

Concrete durability can be defined as the ability to resist through time to the attack of environmental, physical and chemical aggressive conditions. A concrete structure should then be able to maintain the expected performance during its service life. The real issue about concrete durability is related to the intrinsic properties of this material. It presents a higher permeability that allows water and other aggressive elements to enter, leading to carbonation and chloride ion attack resulting in corrosion problems. Thus concrete durability means above all minimizing the possibility of aggressive elements from entering into the concrete,

F. P. Torgal and S. Jalali, *Eco-efficient Construction and Building Materials*,
DOI: 10.1007/978-0-85729-892-8_9,

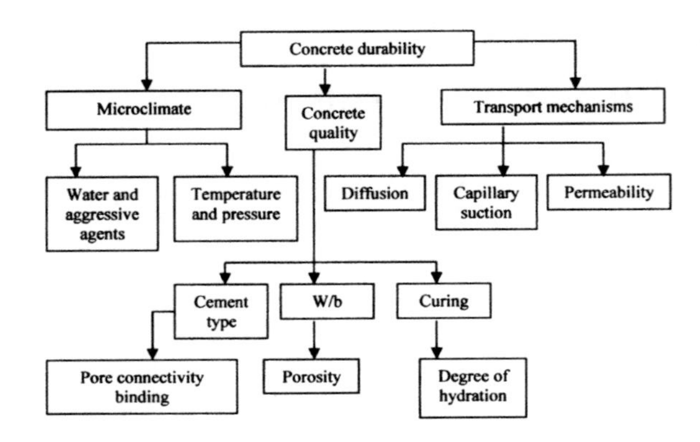

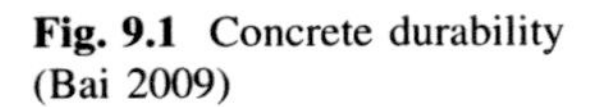
Fig. 9.1 Concrete durability (Bai 2009)

under certain environmental conditions for any of the following transport mechanisms: permeability, diffusion or capillarity (Fig. 9.1).

Avoiding aggressive substances from entering into the concrete is a crucial factor concerning concrete durability, it thus follows that preventing cracking occurrence is of paramount importance to achieve a high durability. Figure 9.2 summarizes the main causes responsible for the occurrence of cracking in concrete.

The degradation mechanisms of concrete can be physical, chemical or a combination of both (Table 9.1).

Due to its importance the following causes that are responsible for the degradation of concrete should be highlighted:

9.2.1.1 Alkali-Aggregate Reaction

The alkali-aggregate reaction (AAR) is a chemical process that occurs between the minerals present in some aggregates, the alkali ions (Na^+ and K^+) and the hydroxyl $(OH)^-$ dissolved in the concrete pore solution. The hydroxyl ions may come from the cement, the mixing water from the aggregates and even from the pozzolanic additions. The reactions developed between the expansive aggregates and Portland cement can be divided into three types depending on the type of aggregates (Reis and Silva 1997; Santos Silva 2005):

(a) Reaction of alkaline hydroxyl with magnesium carbonate of certain limestones, also called alkali-carbonate reaction (ACR).
(b) Reaction between alkali and hydroxyl ions and the amorphous silica in the aggregates, also known as alkali-silica reaction (ASR).
(c) Alkali-silicate reaction, which is identical to ASR but is much slower and takes place not between the silica of the aggregates but among some silicates present in feldspar, it also occurs in certain sedimentary rocks (greywacke), metamorphic (quartzite) and igneous rocks (granites).

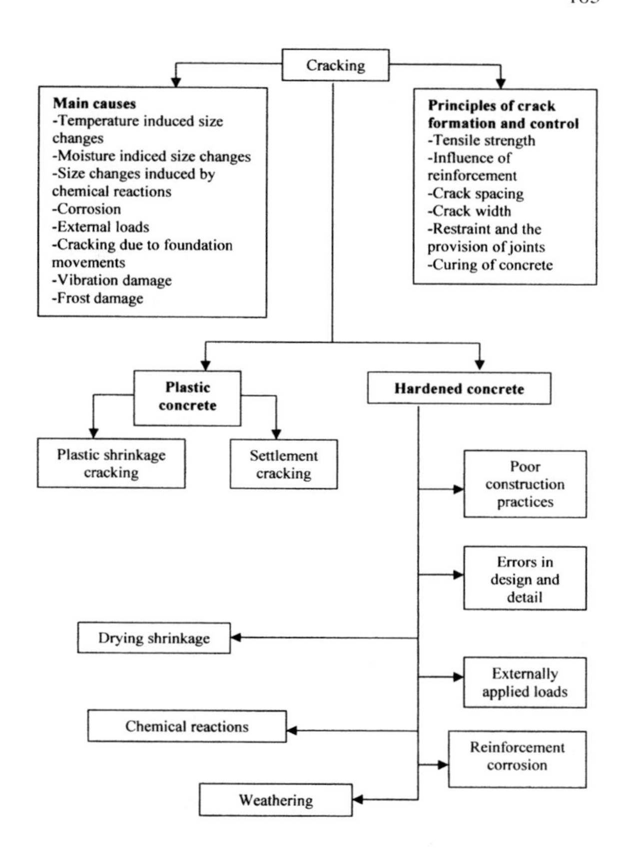

Fig. 9.2 Causes for concrete cracking (Bai 2009)

In the ACR (the first case) the alkali ions from cement will react with the limestone causing the release of magnesium thus exposing the clay phase to water penetration, leading to an expansion phenomenon (Wood and Johnson 1993; Poitvin, 1999). Recently, Grattan-Bellew et al. (2010) showed that this reaction is identical to ASR. In the second case (ASR) which is the most frequent and was mentioned for the first time by Stanton (1940), it encompasses the attack of the siliceous material by alkali hydroxides derived from alkali ions in cement. The occurrence of ASR requires the simultaneous contribution of three factors: (a) a sufficient amount of amorphous silica, (b) alkali ions, (c) water (Hobbs 1988; Jensen 1993; Sims and Brown 1998). This reaction begins with the attack of reactive silica aggregate by the alkalis of cement forming a gel that attracts water by osmosis or diffusion, leading to a volume increase. Since this gel is confined by the cement paste the internal pressure causes the cracking of concrete (Fig. 9.3). This hypothesis was confirmed by several experiments, which showed the occurrence of 4 MPa osmotic pressures. Such tensions are higher than the tensile strength of concrete, so it is conceivable that the gel can be responsible for concrete cracking.

Table 9.1 Causes for concrete degradation (Sarja and Vesikar 1996)

Causes	Processes	Degradation
Static loads	Deformation	Deflection, cracking and rupture
Cyclic loads	Fatigue, deformation	Deflection, cracking and rupture
Biologic		
Micro-organisms	Acid production	Leaching
Bacteria	Acid production	Leaching
Pure water	Leaching	Concrete breakdown
Acid	Leaching	Concrete breakdown
Acids and acidic gases	Neutralization	Steel depassivation
Carbon dioxide	Carbonation	Steel depassivation
Chloride	Ingress, destruction of steel depassivation	Steel depassivation
Steel depassivation + H_2O + O_2	Steel corrosion	Steel expansion, steel section reduction, loss of adhesion
Tension + chloride	Steel corrosion	Prestressing
Sulphate	Crystal pressure	Concrete breakdown
Aggregate (sílica) + álkali	Sílica reaction	Expantion, breakdown
Aggregate (carbonate) + alkali	Carbonate reaction	Expantion, breakdown
Temperature change	Expansion/contraction	Restrained deformation
Humidity change	Shrinkage and expansion	Restrained deformation
Low temperature + water	Icing	Concrete breakdown
Thaw salt + frost	Heat transfer	Concrete scaling
Ice (sea)	Abrasion	Spalling, cracking
Traffic	Abrasion	Wear and rupture
Running water	Erosion	Superficial damage
Turbulent water	Cavitation	Cavities
Electricity	Corrosion	Steel section reduction, loss of adhesion
Magnetism	Corrosion	Steel section reduction, loss of adhesion

Fig. 9.3 Concrete cracking due to ASR

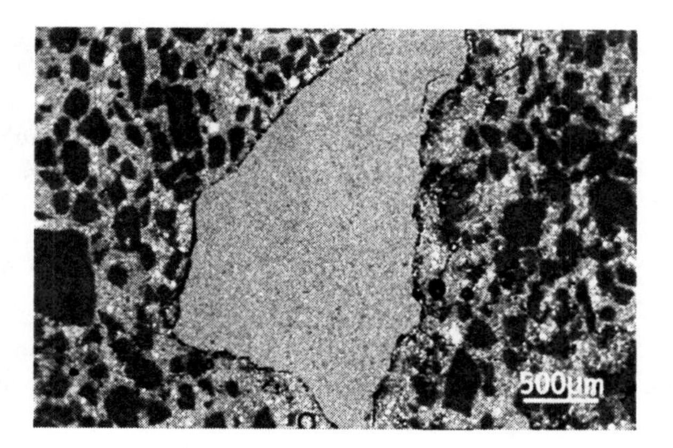

Fig. 9.4 SEM image of an aggregate particle typical of an attack by DEF

9.2.1.2 Sulphate Attack

The attack of sulphates in concrete is characterized by a chemical reaction between the sulphate ions, as an aggressive substance, and the aluminum of the aggregates or with the tricalcium aluminate (C_3A) of the hardened cement paste in the presence of water, forming calcium sulfoaluminate hydrate also known as secondary ettringite and gypsum (calcium sulfate). Secondary ettringite and gypsum are both reaction products that occupied a volume superior to its components which leads to expansion and cracking of concrete. The secondary ettringite is different from the non-expansive form generated during the cement hydration). The sulphate attack occurs when sulphate ions present in ground water, seawater or industrial effluents enter into the pore system of concrete. Concretes with Type I cements are the most vulnerable to this type of attack, on the other hand the most resistant ones are based on pozzolanic cements, although these additions cannot guarantee that the concrete is immune from this type of attack in all situations. All kinds of sulphates lead to concrete degradation, the mechanism and the degree of attack depends on the type of sulfate, being that the magnesium sulphate has a more devastating effect than the calcium or sodium sulphates. A particular type of attack by sulphates is designated by delayed ettringite formation-DEF also known as internal attack by sulphates. This type of attack usually occurs in concretes cured with a heat treatment. DEF can appear when the following conditions are met:

- Absence of external sources of sulphates;
- Heat curing;
- Presence of pores around the aggregates (Fig. 9.4).

The attack by sulphates with the formation of thaumasite is a particular case that unlike the aforementioned two cases, involves the simultaneous formation of gypsum and secondary ettringite. In this particular case the CSH are the ones subject to the attack, not the calcium aluminates hydrates. The replacement of CSH with thaumasite (a soft paste) leads to a strength loss of the concrete. This type of attack is much more severe than those involving cracking expansion because it reduces the concrete strength in a drastic manner. This type of attack needs sulphate and carbonate ions, CSH and water. Since limestone aggregates are a source of carbonate ions, concretes made with limestone aggregates that are used in foundations and in contact with sulphate solution are subject to a high risk of attack.

9.2.1.3 Carbonation

Concrete carbonation is a process by which atmospheric carbon dioxide reacts with the cement hydration products to form calcium carbonate. The importance of this phenomenon relates to the fact that it reduces the alkalinity of the concrete to a pH of almost 8. Since the steel passivation layer (an iron oxide layer that covers steel reinforced bars), needs a high pH between 12 and 14, the carbonation phenomenon can be responsible for the steel depassivation. The main factor that controls the carbonation process is the diffusivity of the cement paste, which in turn is a function of the porous network, hence the W/B ratio is a key factor in determining the carbonation depth. For instance, in concrete with a W/B = 0.6 a carbonation depth of 15 mm can be achieved after 15 years, but if the concrete has a lower W/B = 0.45, the same carbonation depth will take 100 years to reach (Wiering 1984).

9.2.1.4 Chloride Penetration

The ingress of chlorides into the concrete leads to the corrosion of steel reinforcement and the cracking of the concrete caused by the swelling of the oxide corrosion products. The ingress of chlorides into the concrete can occur by using contaminated aggregates, by mixing water or by the use of admixtures containing chlorides ions. Not all chloride ions contribute to the corrosion of the steel reinforcement, some react with the hydration products of the cement being locked, only the free chloride ions can initiate the corrosion process.

9.2.1.5 Corrosion of Steel Reinforcement

The corrosion process begins when the loss of the passivation layer occurs or when a certain critical value of chloride ions enters into the concrete. The steel corrosion occurs due to an electrochemical action, when metals of different natures are in electrical contact in the presence of water and oxygen. The process consists in the anodic dissolution of iron (Fe):

$$\mathrm{Fe} \rightarrow \mathrm{F}^{2+} + 2\mathrm{e}^{-} \text{ (anode)}$$

Fig. 9.5 Spalling of a concrete cover due to the expansion of corrosion products

when the positively charged iron ions pass into the solution and the excess of negatively charged electrons goes to steel through the cathode, where they are absorbed by the electrolyte constituents to form hydroxyl ions $(OH)^-$

$$\frac{1}{2}O_2 + H_2O + 2e^- \rightarrow 2(OH)^- \text{ (cathode)}$$

These in turn combine with the iron ions to form ferric hydroxide, which then converts into rust:

$$F^{2+} + 2(OH)^- \rightarrow Fe(OH)_2$$

The volume increase associated with the formation of the corrosion products will lead to cracking and spalling of the concrete cover (Fig. 9.5).

There are basically two ways to control the durability of concrete. One is by imposing limits to the composition in terms of the W/B ratio, minimum amount of cement, by guaranteeing a minimum concrete cover or a minimum compressive strength class. This is an easy approach, but a simplistic and conservative one. Another variant is linked to the performance requirements and has the advantage of allowing for design optimization reaching cost benefits, without sacrificing the safety of the structure (Ferreira 2009). In this second approach, the concrete performance is specified but not the way in which to be achieved. However, the transition from prescriptive standards to performance standards is not easy because it requires the development of performance tests that evaluate the materials' performance on the site conditions. In the new regulations the concrete durability

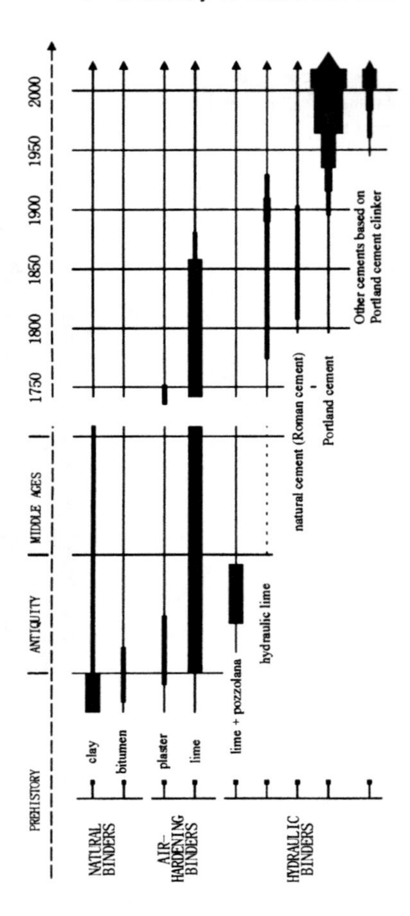

Fig. 9.6 Use of binders during history (Elsen et al. 2010)

requirements are defined in terms of composition and property thresholds. They may also be defined in terms of performance-based specifications.

9.2.2 Renders Used in Ancient Buildings

Throughout history renders containing different binders were used (Fig. 9.6). Renders of buildings built before the XX century are composed mostly of air hardening binders and/or hydraulic additions. Its execution although based on a very skilled workforce was very time consuming, the final result was a material with a low compressive strength and sometimes with low durability. In the twenty-first century, the appearance of Portland cement, associated with high early compressive strength led to a generalization of blended mortars based on aerial lime-Portland cement and even of mortars containing Portland cement as the sole binder. The use of conservation mortars based on Portland cement for the

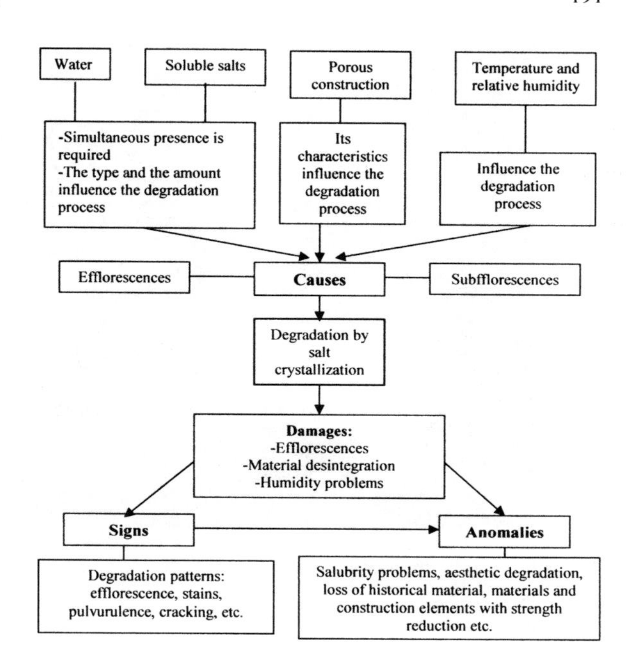

Fig. 9.7 Render degradation by crystallization of soluble salts (Gonçalves and Rodrigues 2010)

rehabilitation of ancient buildings led to the development of several pathologies: They are chemically incompatible with lime-based mortars; they are responsible for the introduction of soluble salts; have a low permeability and a high modulus of elasticity that is unable to accommodate for masonry deformations. The efflorescences are actually one of the most serious pathologies associated with almost all the renders of ancient buildings. The solution to this problem can hardly pass by the use of the so-called "miracle" products commercially available (Lubelli et al. 2006a, b). Gonçalves and Rodrigues (2010) also reported that the degradation by crystallization of soluble salts is a major mechanism of the degradation of mortars in ancient buildings. These authors present a scheme of this pathology (Fig. 9.7). The crystallization of soluble salts may occur on the render surface (efflorescence) or on the interface render/masonry (sub efflorescence). The latter case is more serious because it leads to a faster render deterioration (Groot et al. 2009).

Very often the degradation of renders by crystallization of soluble salts gets worse after a rehabilitation operation. The explanation is related to the use of renders with low water vapour permeability (Gonçalves 2007; Gonçalves et al. 2008) as it happens in Fig. 9.8 which presents the case of a wall loaded with sodium chloride salts from a flood that was rehabilitated with a new render.

The durability of the renders of ancient buildings is dependent on the rehabilitation operations that are able to identify the original renders and their compositions, to ensure an adequate rehabilitation. This subject will be addressed in Sect. 9.5. The worst pathologies associated with gypsum plasters include moisture or water leaks, which result either from water pipe ruptures or damaged roofs (Fig. 9.9).

Fig. 9.8 Deterioration of a new render between 1991 and 1995 in an ancient building in The Netherlands

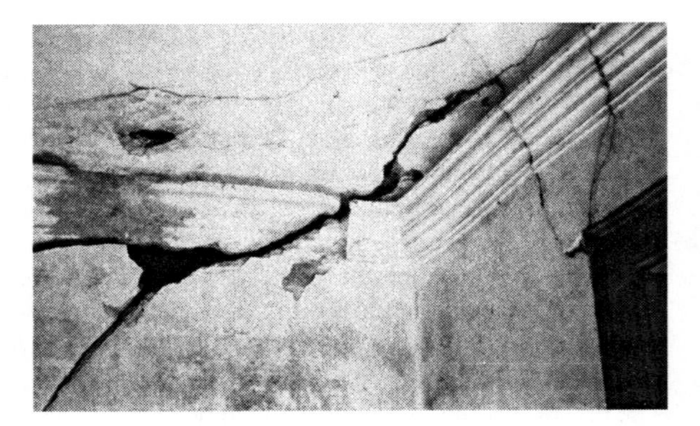

Fig. 9.9 Pathology of gypsum plaster due to the degradation of the building roof (Pacheco-Torgal and Jalali 2009b)

This behavior is worsened by the fact that gypsum plaster gets softer and dissolves when exposed to water. The use of waterproof paint also contributes to the appearance of stains and ultimately to the disintegration of gypsum plaster, because they prevent the water vapour permeability of these walls leading to the appearance of subfflorescences. Although less frequent, the use of mortars based on Portland cement for the rehabilitation of gypsum plasters could result in loss of adhesion due to the formation of thaumasite (Corinaldesi et al. 2003). Less serious but not less often appear the pathologies associated with the cracking of gypsum that are caused by structural problems of the building, including pavement deformations that affect the behavior of the walls. Another common pathology in gypsum plasters, has to do with the fall of sections of decorative pieces. This occurrence is caused by the loss of adhesion between the decorative piece and the

Fig. 9.10 Falling parts in decorative gypsum plaster piece (Pacheco-Torgal and Jalali 2009b)

gypsum plaster. The reason lies in the movements of contraction and expansion of the wood lath and also to the presence of moisture that reduces the mechanical strength of the plaster materials (Fig. 9.10).

9.3 Concrete Conservation and Retrofitting

9.3.1 Measures to Minimize the Occurrence of ASR in Concrete

Given the key role of water to initiate ASR, the use of concrete waterproof coatings in order to prevent its access into the concrete, could minimize the occurrence of ASR. Santos Silva (2005) mentioned that the use of pozzolans in concrete is the most effective way to substantially reduce ASR in concrete containing reactive aggregates. Since they consume nearly all the calcium hydroxide generated during the hydration of cement there will not be enough calcium hydroxide to initiate the harmful reaction. The same author also mentioned that metakaolin is more effective than fly ash in the inhibition of ASR, because it is a more reactive pozzolan, consuming more calcium hydroxide. He reported the need for replacing 20% of cement with fly ash to prevent ASR, and just 15% of metakaolin to achieve the same effect. Moisson (2005) suggested the use of crushed reactive aggregates as additives to reduce the occurrence of ASR in the concrete. Other authors (Charles-Gibergues et al. 2008) also confirm that ASR phenomenon could be prevented by using ground reactive aggregates in the concrete mix. This additive shows pozzolanic behaviour, but its activity depends on its fineness and on the petrographic characteristics of the reactive aggregates. Ichikawa (2009) mentioned that the efficacy of pozzolans in the suppression of ASR is dependent on its calcium content. The use of rich calcium fly ash can worsen this problem. Sousa et al. (2010) reported that replacing 30% of Portland cement with tungsten mine waste provides an effective solution to mitigate ASR.

9.3.2 Concrete Surface Treatments

The use of concrete surface treatments with waterproofing materials to prevent the access of aggressive substances, is a way of contributing to the increase of concrete durability. The most common surface treatments use polymeric resins based on epoxy, silicone (siloxane), acrylics, polyurethanes or polymethacrylates. Bijen (2000) mentioned that epoxy resins have low resistance to ultraviolet radiation and polyurethanes are sensitive to high alkalinity environments. On the other hand although some waterproof materials are effective for a particular transport mechanism (diffusion, capillarity, permeability) they aren't for another. Moreira (2006) compared the waterproofing capacity of concrete with three polymeric resins (epoxy, silicone, acrylic) and mentioned that the silicone based is more effective (99.2%) in reducing water absorption by capillarity than the epoxy resin (93.6%), but in terms of chloride diffusion the epoxy resin is 100% effective, while the silicone varnish does not go beyond 67.5%. Medeiros and Helene (2008) used a water repellent material based on silane-siloxane noting that although it is effective to reduce the water absorption by capillarity of concrete (reduced from 2 to 7 times), it only managed to achieve a reduction of the chloride diffusion from 11% to 17% and also failed to prevent the access of water by permeability. Pacheco-Torgal and Jalali (2009a) confirm that the surface treatment of concrete with a water repellent material is effective, but above all more cost-effective when compared with the alternative of using a polymer additive in the composition of concrete. More recently, Zhang et al. (2010) mentioned that the use of geopolymers for sealing the concrete surface can be a more effective technique than the organic polymers used so far. A new technique for waterproofing of the concrete surface uses bacteria, which is responsible for the production of calcium carbonate crystals, reducing the porosity and the permeability of concrete. In terms of sustainability, this new technique is more advantageous than the current use of polymers (organic or inorganic). De Muynck et al. (2008b) mentioned that the use of the bacteria *Bacillus sphaericus* led to a reduction from 65% to 90% of water absorption by immersion and a reduction from 10 to 40% in the diffusion of chlorides. The investigations of De Muynck et al. (2008a) show that in terms of water absorption by capillarity and water permeability the performance of bacteria treatments is identical to waterproof materials based on organic polymers. Jonkers et al. (2010) studied the incorporation of bacterias inside the concrete during the mixture phase to assess the possibility that these might help to fill the cracks. The results although promising show that they only stay active at early ages and that as the cement hydration proceeds they disappear. Tittelboom et al. (2010) also obtained promising results in terms of sealing concrete cracks with the help of bacteria, however, they mentioned that this is possible only for cracks up to 10 mm in depth and that it is necessary to protect the bacteria with silica gel from the high alkalinity of the concrete.

Table 9.2 Electrochemical protection and repair techniques (Da Silva 2007)

	Cathodic protection	Desalinization	Increase of pH
Purpose	Steel rebar polarization	Chloride removal repassivation	Increase of pH/ repassivation
Polarization time	Permanent	6–10 weeks	1–2 weeks
Current density (m^2 of concrete)	3 a 20 A/m^2	0.8 a 2 A/m^2	0.8 a 2 A/m^2

9.3.3 Electrochemical Techniques to Prevent or Reduce Steel Corrosion

The use of electrochemical techniques to prevent or even to reduce the corrosion of steel reinforcement in concrete structures, use an electric current between the steel rebars and an external element (anode). This makes the steel rebars to act as a cathode (electronegative charge), thus forcing the chloride ions negatively charged away from the rebars. Apart from the removal of chloride ions it may happen that hydroxide ions will also form, leading to the restoration of the passivation layer of the steel rebars. According to the Federal Highway Administration-FHWA the cathodic protection is the only technique to prevent corrosion on bridges (submitted to an almost constant salt exposition to prevent ice formation), irrespective of the chloride content in concrete (Bijen 2000). Other authors mentioned some side effects associated with cathodic protection such as: the use of high electric current can reduce the adhesion between the steel rebars and the concrete; hydrogen embrittlement of high strength steels and the increasing of the alkalinity level can be responsible for the occurrence of ASR if the concrete contains reactive aggregates (Pedeferri 1996). Da Silva (2007) describes three techniques for electrochemical protection and repair, which differ in the intensity of the electric current and the polarization time (Table 9.2).

Betolini et al. (2008) analyzed the effects related to the use of cathodic protection in reinforced slabs for five years mentioning that the use of a current density above 1.7 mA/m^2 prevented steel corrosion. Da Silva (2007) studied the economics of repairing an on-shore concrete structure with cathodic protection versus the alternative of using stainless steel reinforcement, concluding that although the latter have a higher cost (5 to 7 times higher than ordinary steel) it is the best cost effective option. It is worth mentioning that this study only examined replacement rates of ordinary steel with stainless steel below 50%; the use of higher replacement rates increases the economic competitiveness of cathodic protection. Lambert (2009) studied the economics of cathodic protection mentioning that it can cost less than 5% compared to the option of full replacement of the entire concrete structure.

Table 9.3 Structural compatibility (Morgan 1996)

Properties	Relation between the repair mortar (Rp) and the concrete substrate (Cs)
Strength in compression, tension and flexure	Rp ≥ Cs
Modulus in compression, tension and flexure	Rp ~ Cs
Poisson's ratio	Dependent on modulus and type of repair
Coefficient of thermal expansion	Rp ~ Cs
Adhesion in tension and in shear	Rp ≥ Cs
Curing and long term shrinkage	Rp ≥ Cs
Strain capacity	Rp ≥ Cs
Creep	Dependent on whether creep causes desirable or undesirable effects
Fatigue performance	Rp ≥ Cs

General requisites for repair mortars

9.3.4 Mortars for Concrete Structure Repairs

The retrofitting of a given concrete structure using repair mortars requires that the compliance with requirements for structural stability has been previously addressed. This means that this operation is done in structures without rebar's section loss, hence there is no need for specific strengthening operations. Nor does it have problems related to ASR that cannot be solved using repair mortars. The application of the repair mortar is preceded by cleaning operations of the concrete surface to remove pieces of degraded concrete. Moreover, as the roughness of the concrete substrate affects the performance of most repair mortars, it becomes necessary to artificially increase its roughness, regardless of the cleaning operation. Generally, this operation is carried out using mechanical devices such as:

- Brushing with wire brush;
- Poking with jackhammer;
- Scraping with pneumatic tool;
- Smoothing with multiple disks;
- Projection of grit;
- Sandblasting;
- Water jet.

In order to ensure structural compatibility with the concrete substrate, repair mortars must meet the requirements defined in Table 9.3.

Figure 9.11 shows the disadvantages related to the use of materials with different modulus of elasticity.

The repair mortars must also comply with the requirements defined in Fig. 9.12

Another fundamental property of repair mortars is a rapid adhesion to the concrete substrate, allowing that the structure is back into service as soon as possible. Several tests like the "*pull-off*" and the "*slant shear*" have been used to

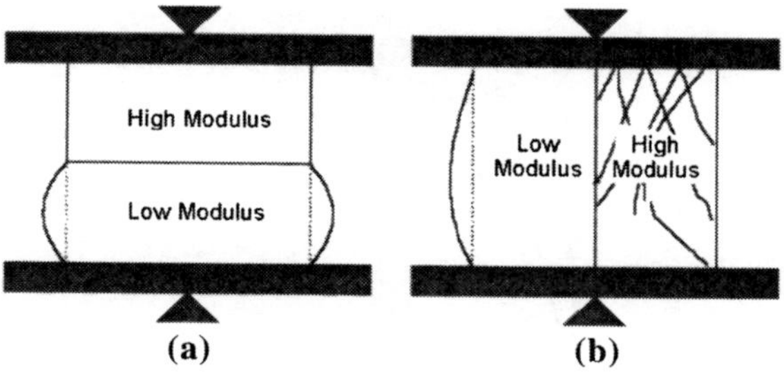

Fig. 9.11 Mechanical behavior of materials with different modulus of elasticity. **a** Load perpendicular to the interface, **b** Load parallel to the interface

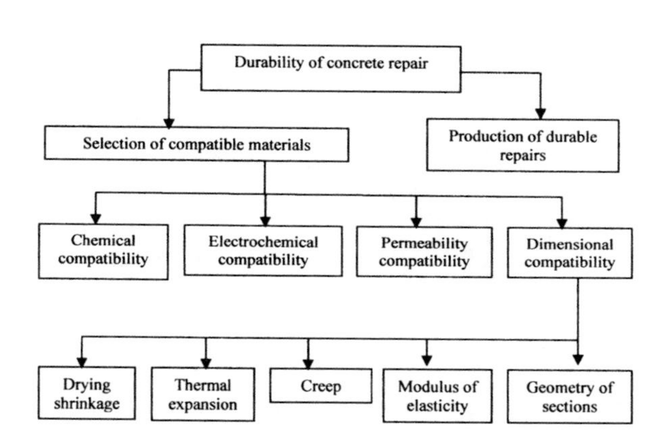

Fig. 9.12 Factors that influence the durability of repair mortars (Morgan 1996)

quantify the adhesion strength (Austin et al. 1999; Momayez et al. 2005). Currently, most repair mortars fall into two categories, mortars based on organic binders (epoxy resin or polyester) or those based on inorganic binders like Portland cement. The latter are more cost effective and less toxic, being available as a commercially pre-pack mixture of Portland cement, aggregates, silica fume, fibers and other additives. More recently investigations about geopolymers reveal a third category of mortars with high potential to be used in the field of concrete structures retrofitting (Pacheco-Torgal et al. 2008). Since the adhesion to the concrete substrate is a crucial property of repair mortars, some results related to the comparison between geopolymer mortars and commercial products for the repair of concrete structures are presented below. The adhesion strength was evaluated using the slant shear test. The slant shear test uses square prisms made of two halves, one of concrete substrate and another one of repair material, tested under axial compression. The adopted geometry for the slant shear specimens was a $50 \times 50 \times 125\ mm^3$ prism with an interface line at 30° to the vertical. Bond strength was calculated by dividing the maximum load at failure by the bond area and was obtained from an average of four specimens determined at the ages of 1, 3, 7 and 28 days of curing. In order to increase the specific surface of the concrete substrate, an etching procedure was carried out. The concrete surface was immersed in a 5% hydrochloric acid (HCl) solution for 5 min and then carefully washed to ensure the removal of $CaCl_2$ which results from the reaction between HCl and $Ca(OH)_2$. The specimens were

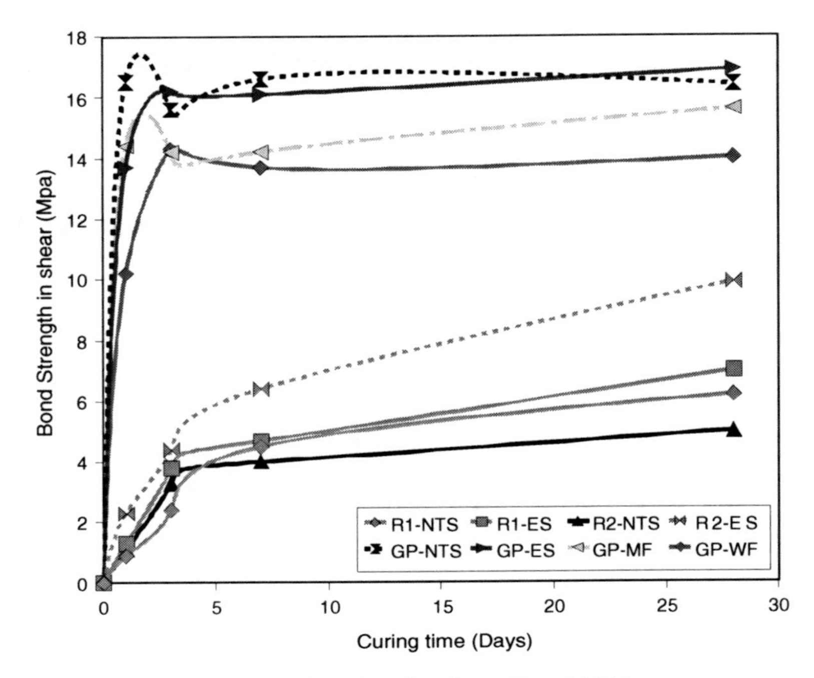

Fig. 9.13 Adhesion strength using the "slant-shear" test (Torgal 2007)

named after the repair materials and concrete substrate surface treatments. Specimens using concrete substrate repaired with commercial product R1 with and with no surface treatment were named respectively, R1-ES (etched surface) and R1-NTS (no treatment surface). Similarly, when the geopolymer-based binder was used to bond the two halves they were named GP–ES and GP–NTS, respectively. Slant specimens with substrate surface treatment as cast against metallic formwork, and as cast against wood formwork were also used repaired with geopolymeric binder and were named, GP-MF and GP-WF, respectively. The results of the effect of the several repair solutions on the average adhesion strength are shown in Fig. 9.13. It can be seen that the specimens repaired with the geopolymeric mortar present the high adhesion strength′s even at early ages. Specimens repaired with geopolymeric mortar with 1 day of curing have higher bond strength than specimens repaired with current commercial products after 28 days curing. Specimens repaired with the geopolymeric mortar appear not to be influenced by the chemical treatment in sawn concrete surface substrates, but by the use of concrete surfaces as cast against formwork. Those kinds of surfaces are rich in calcium hydroxide but lack exposed coarse aggregates which could contribute toward improving the bond strength due to silica dissolution from the aggregate surface. The strength performance of commercial repair products is very dependent on the curing time and this constitutes a serious setback when early bond strength is required. The results show that bond strength using repair product R2 is clearly influenced by the surface treatment.

In order to evaluate the economic profitability of the different repair solutions, comparisons between the costs of materials were made. In Table 9.4, the cost of the raw materials used to prepare the repair mortars and also the cost of the repair mortars are reported.

Even if the current commercial repair materials had the same mechanical performance as that of geopolymeric mortars, the cost of the cheapest one (R1) is still 6.9 times higher than the geopolymeric mortar. When comparing the cost to the adhesion strength ratio the differences are even higher, with the cost of the cheapest solution with current commercial repair products (R1-ES) being 13.8 times higher than the solution with the geopolymeric mortar (Fig. 9.14).

9.4 Render Rehabilitation

9.4.1 Materials Characterization and Design of Repair Mortars

The rehabilitation of renders in ancient buildings should be preceded by an analysis of the depth of degradation and the historical or artistic value of the building. In case of buildings with historical relevance the first option should be to preserve the original renders through maintenance operations, if that is not possible then a rehabilitation operation should be carried out. If the level of degradation is very high a partial render replacement must be enforced. The total replacement must always be the last operation to consider. Any work must be preceded by an inspection of the walls to be rehabilitated, because since the walls of ancient buildings are rather irregular they will need a high render thickness, which constitutes an important technical subject to deal with. Beyond the historical value implicit in the use of original materials and techniques the compatibility between the rehabilitation mortars and the materials to be rehabilitated represents a crucial aspect to be met. Materials with different permeabilities, different modulus of elasticity, different adhesion strength, different absorption of water or even with or without vegetable or animal fibres, can hardly constitute good solutions for rehabilitation purposes, because sooner or later they will be associated with the appearance of pathologies. The simple fact of changing the particle size of a mortar is enough to alter its performance (Genestar and Pons 2003), because coarser aggregates lead to higher mechanical strength, lower porosity and lower capillary absorption. This simple fact allows us to understand the high number of variables that may contribute to the poor performance of a rehabilitation solutions. The first step related to the rehabilitation of renders in ancient buildings is to find out the composition of the original renders. The identification of the aggregates is simple (Fig. 9.15).

In order to to separate the silica aggregates from the binder, the lime mortars are usually dissolved in chloric acid (1:3). The insoluble residue is then weighed and

Table 9.4 Cost of repair mortars

Repair solution	Materials cost (€/ton)							Total cost	
	Geop. mortar (25 €/ton)	Aggregates (7.7 €/ton)	Waterglass solution (428 €/ton)	Sodium hydroxide (728 €/ton)	Calcium hydroxide (100 €/ton)	Repair material R1 (910 €/ton)	Repair material R2 (1646 €/ton)	(€/ton)	(€/m^3)
Geop. mortar	6.3	3.3	85.6	28.5	2.8	–	–	126.5	263
Material R1	–	–	–	–	–	910	–	910	1820
Material R2	–	–	–	–	–	–	1646	1646	3292

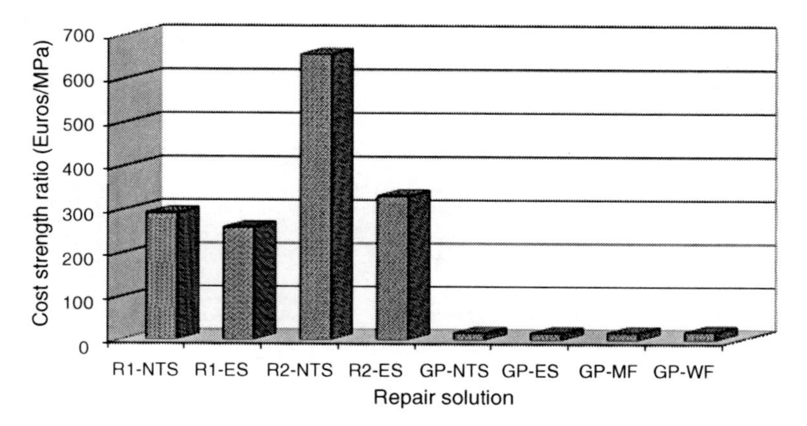

Fig. 9.14 Cost to adhesion strength ratio after 28 days curing according to repair solution

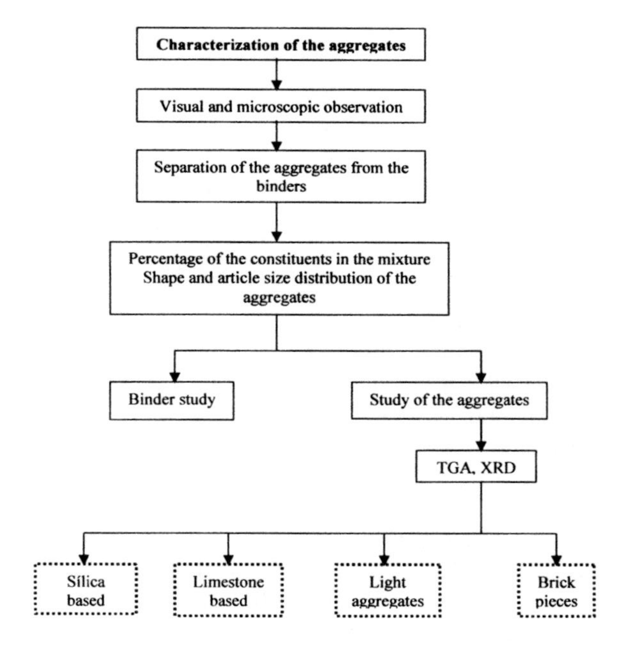

Fig. 9.15 Characterization of the aggregates present in original mortars (Marques 2005)

passed through a set of sieves to determine the particle size of the sand (Adriano et al. 2007). Regarding the characterization of the binders (gypsun, lime and Portland cement), the methodology is more complex (Fig. 9.16) and given the numerous chemical reactions that can take place, the results are not always absolute. The methodology is relative because other authors were able to distinguish gypsum and anhydrite compounds using FTIR tests (Carbo et al. 1996).

The gypsum binders can be identified by TGA because they present a small peak in the range 120°C to 200°C which corresponds to the absorbed water or even

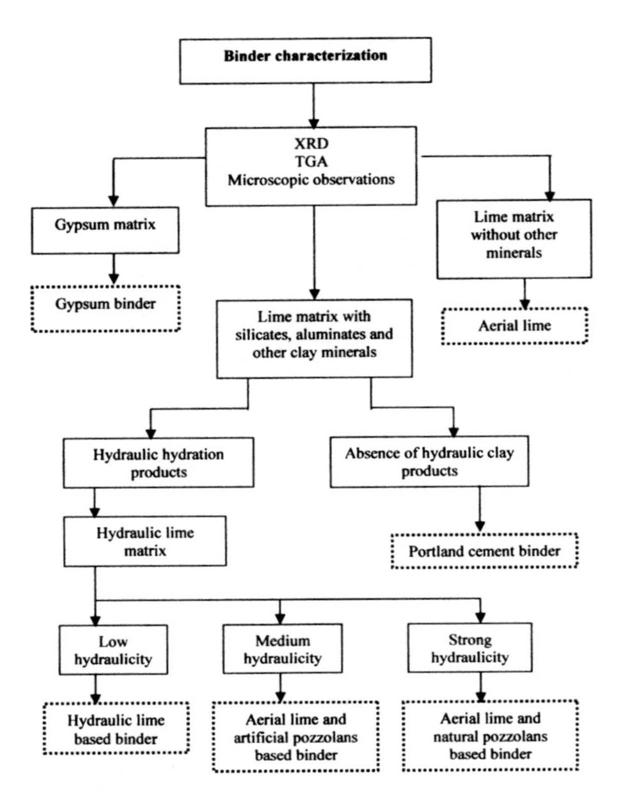

Fig. 9.16 Characterization of the binders present in original mortars (Marques 2005)

by XRD in order to confirm the presence of calcium sulfate peaks. The hydraulicity of a mortar can be obtained using TGA to assess the loss of structural water between 200°C and 600°C and the loss of carbon dioxide between 200 and 600°C. The hydraulicity ratio (Hr) can be defined as the quotient between weight loss attributed to CO_2 and to water decomposition (Ugurlu and Boke 2008). Aerial limes have a CO_2/H_2O_{st} ratio higher than 10. Lime-pozzolan binders have a ratio below 10. Marques et al. (2006) mentioned that a CO_2/H_2O_{st} ratio lower than 3 corresponds to a strong hydraulic character (natural pozzolans or Portland cement), whereas binders with a ratio between 3 and 6 have a medium hydraulic character (lime and artificial pozzolans like crushed bricks). This classification is not valid when limestone aggregates are used because they increase the CO_2 emissions. The characterization of the physical and mechanical properties of the renders of ancient buildings involves not only laboratory tests but also on-site tests. These tests can be subdivided according to their degree of intrusion:

- Non destructive tests: low permeability using Carsten tubes (Fig. 9.17c), dynamic modulus of elasticity;
- Semi-destructive and destructive tests: assessment of the deformation resistance using the ball impact test (Fig. 9.17a), the controlled penetration test (Fig. 9.17b) and the pull-off test to assess the adhesion strength (Fig. 9.17d).

Fig. 9.17 On-site tests to assess the performance of mortars: **a** ball impact test, **b** controlled penetration test, **c** Carsten tube (Veiga 2003), **d** pull-off test (Flores-Colen et al. 2007)

Ball impact test—this test uses a 50 mm diameter sphere to make an impact in the render. The crash strength is assessed from the dent diameter caused by the ball shock.

Controlled penetration test— it involves the penetration of a steel nail with several shots, after each shot the penetration is measured.

Carsten tube— it uses a set of tubes filled with water which are attached to the wall, after a certain time the amount of water absorbed by the wall is measured.

Pull-off test—this test evaluates the force required to pullout a render specimen.

After the analysis of the composition of the original renders as well as their physical and mechanical properties, it is necessary to design the characteristics of the rehabilitation mortars and to assess the compatibility between the two mortars. An experimental methodology for that purpose is presented in Fig. 9.18.

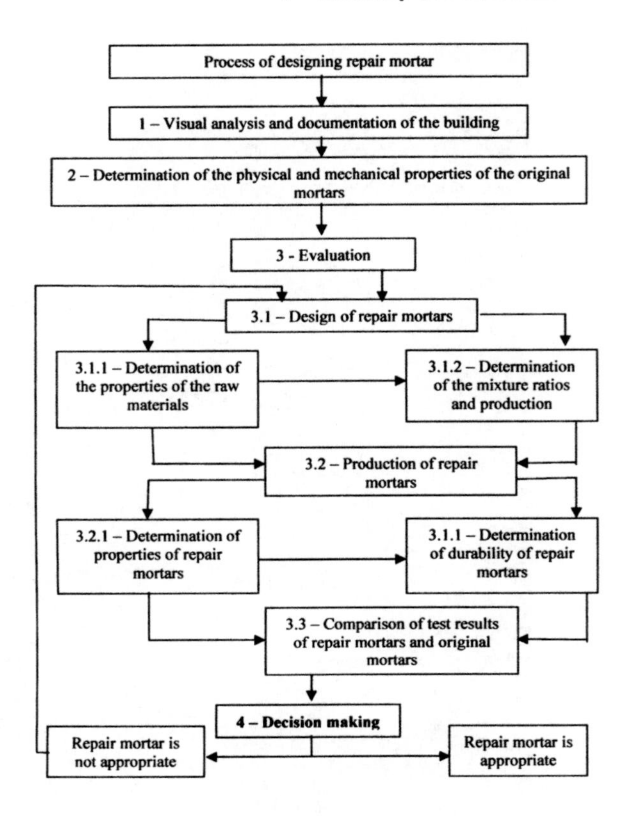

Fig. 9.18 Experimental method used in the process of designing repair mortars (Arioglu and Acun 2006)

9.4.2 Rehabilitation of Gypsum Plasters

This kind of work requires in most cases the structural rehabilitation of the building elements (walls and ceilings), whose malfunction led to the pathology of the gypsum plasters. The recovery of the structural integrity of the wooden ceilings is in most cases a necessary step in these works, whether using fibre reinforced polymers-FRP applied to its back or as a reinforcement of the wood beams, followed by the reapplication of the wood lath. In the walls and ceilings where a redo of the wood structure was required, after placing the lath, comes the application of the gypsum plaster tightened by hand in a direction perpendicular to the lath. In some cases where the ingress of moisture favored the development of algae and mosses, it is first necessary to carry out its removal (Cotrim et al. 2008). Some authors suggest that the use of a methylene chloride solvent-based stripper causes the expansion of the ink allowing for its mechanical removal. The existence of previous rehabilitation mortars containing Portland cement must also be removed. The same should be done to the paintings that have been implemented without the respect for compatibility and permeability with the gypsum plasters. The paint removal must be done by scraping with brushes to avoid damaging the gypsum

Fig. 9.19 Portuguese fortress of "Nossa Senhora da Conceição" in the Persian Gulf

plaster, but if the mechanical operation is not enough then a chemical stripper can be used. The traditional strippers used for cleaning paint are completely inadvisable because they introduce harmful salts and leave the plaster surface with bites. Regarding the degraded gypsum decorative elements their rehabilitation is dependent on their types. While the frames are run on the ceiling, using a suitable profile made from the profile of the original frames, the center ceiling medallions and other decorative elements are previously executed off-site and only then placed on the ceiling with an adhesive plaster. The execution of decorative gypsum plaster pieces is a work that requires specialized labor because it is a very detailed job that tries to fully reproduce not only the original forms but sometimes also the original pigmentation.

9.4.3 Air Lime Mortars with Vegetable Fat

Historically speaking, the use of air lime mortars with the addition of vegetable fat goes back to Vitruvius of the Roman Empire (Bailey 1932; Albert 1995). In Portugal, the architect Quirino da Fonseca published in the early 1990s a little book (Sá 2002, 2005), where he mentioned the addition of small amounts of vegetable oil during the lime slaking process. He also mentioned that these materials were used by Portuguese masons to built ancient fortresses, including "Nossa Senhora da Conceição", a fortress located in Gerum island, Ormuz, in the Persian Gulf. The construction of that fortress took place in 1507, being constituted by eight outer towers and a central one (Fig. 9.19). In 1873, more than three hundred years after the construction of "Nossa Senhora da Conceição" fortress, A. W. Stiffe a LouTenent of the British navy visited the interior of the fortress having made a description of the conservation status for the Geographical Magazine. He states that "The mortar used was excellent, and much more durable than the stones" (Rowland 2006). This allows us to have a good impression about the durability of those mortars, mainly as most of them were placed near harsh sea conditions. In 1570, the Venetian architect Palladio, mentioned the use of linseed and nut oil to obtain water proof lime-pozzolan mortars (Palladio 1570). Also, Manuel Azevedo Fortes in his book, "The Portuguese Engineer" published in 1729, mentioned the use of olive oil in the air lime slaking process, being that the olive oil should be added to the lime while it is in a boiling state (Sá 2005).

Since the middle of the 1990s, a Portuguese manufacturer has been selling an air lime named "D. Fradique", which is produced with the addition of olive oil waste. The production of this air lime started after the Portuguese architect Quirino da Fonseca was chosen to be in charge of the conservation works of the walls of the Castle S. Jorge in Lisbon. At that time he tried to reproduce the characteristics of ancient lime mortars with the addition of vegetable fat. The firm responsible for the production of "D. Fradique" air lime used a quasi-industrial process. After the calcination of the limestone rocks they are ground in a jaw mill. The air lime slaking process is handmade and at the same time the olive oil wastes are added. Quirino da Fonseca recommends the following proportions to the manufacturer of "D. Fradique" air lime: 25 kg of lime; 1.5 kg of olive oil wastes; 10 l of water. Of course the manufacturer uses other proportions that are under commercial secrecy (Sá 2005). The hardening of the "D. Fradique" air lime occurs by a carbonation mechanism as it happens with other air limes. The mortars made with "D. Fradique" air lime have several advantages over the current air lime-based mortars:

- Higher workability;
- Higher water vapour permeability;
- Lower capillary water absorption;
- higher water proofing behavior;
- Higher resistance to fungal development.

Sá (2005) studied the behavior of several mortars used as renders for stone based masonry, saying that although "D. Fradique" air lime mortars possess high water proof cabability than other mortars, at the same time they have low impact resistance and low adhesion resistance (Table 9.5).

The fact that this lime presents a less porous structure, associated with the vegetable oil fat presence contributes to delay the carbonation process, thus also delaying the mechanical resistance development. Veiga (2003) mentioned that the use of renders based on this particular type of lime may lead to some failures that can be explained by the carbonation delaying effect mentioned above. Vegetable oils are constituted by glycerides (esters of glycerol and fatty acids). The glyceride is not chemically stable in highly alkaline environments like cement mortar. When that happens, the carboxyl group of the fatty acid anion will coordinate stringly with calcium oxide (Justnes et al. 2004). Other authors (Vikan and Justnes 2006) studied the waterproofness performance of hydraulic binders mixed with vegetable oil fat. They mentioned that using just 0.5% of vegetable oil by cement weight allows good mortar performances and also that rapeseed oil is one of the most effective oils for that purpose, even more than olive oil, and it is cheaper. The majority of additives used to enhance water proofing of concrete and mortars (resins and polymers) came from the oil industry. Therefore, the eco-efficiency

Table 9.5 Comparison of the mortar properties (Sá 2005)

Test	Mortars according to the binder used			
	D. Fradique air lime (1:3)	Air lime	Hidraulic lime (1:3)	Air lime and cement (0.5:0.5:3)
Flexural resistance-Rt (MPa)	0.22	0.21	0.20	0.47
Compression resistance-Rc (MPa)	0.65	0.44	1.56	2.67
Rt/Rc	0.34	0.48	0.13	0.17
Impact resistance with a sphere Ø impact (mm)	18.6	20.7	12.6	11.8
Adhesion resistance (MPa)	0.048	0.056	0.057	0.129
Water absorption by Carsten tubes-10 min (cm^3)	0.1	36.1	2.1	3.1
Capillary water absorption (water mass after 5 min) (g)	2.22	23.92	9.66	8.57

context implies that new environmental friendly additives must be investigated. It is not without irony that the past can teach us something in this area. Stolz (2007) mentioned that one of the disadvantages of using mortar renders based on vegetable oils relates to the fact that since UV lights can oxidize the fatty acids, this leads to a reduction, with time, of the water proofing ability near the surface. Cechova et al. (2008) studied mortars with 1% of linseed oil addition. These authors state that oil addition increases the flexural and compressive strength of hydraulic mortars, but for air lime mortars an inverse behavior was observed. Both mortars show a decrease in water absorption but this effect is higher for air lime mortars showing a water absorption 10 times lower. More recent investigations (Chechova et al. 2010) confirm that oil addition delays the hardening process. Also mortar performance is dependent on the oil amount. The use of 1% of linseed oil improves mortar strength but when 3% is used, air lime mortars show a strength reduction. The results also show that oil addition improves the resistance to salt and freeze–thaw cycles for both percentages.

9.4.4 Renders for Salt Laden Masonry Substrates

If for any reason the soluble salts could not be removed the designer must decide the most appropriate render, among five types defined under the European investigation project COMPASS (Groot et al. 2009):

1. Renders for salt transport (slow);
2. Renders for salt transport (fast);
3. Salt accumulation renders;
4. Moisture sealing renders;
5. Renders for salt blocking.

Gonçalves and Rodrigues (2010) present different types of renders for salt laden masonry substrates (Fig. 9.20).

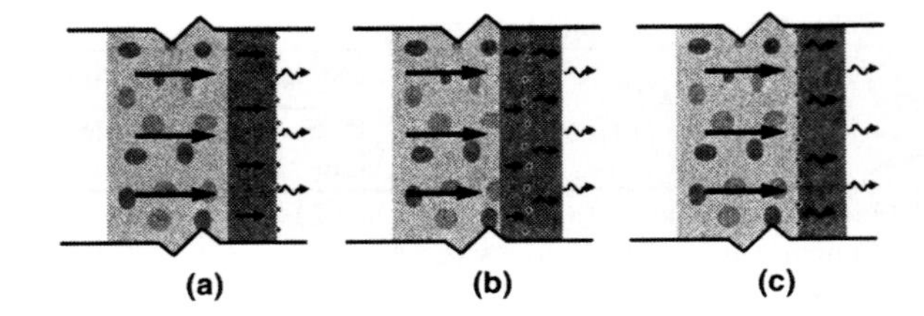

Fig. 9.20 Different renders for salt laden masonry substrates (Gonçalves and Rodrigues 2010) **a** Render for salt transport. **b** Salt accumulating render. **c** Render for salt blocking

The mixture and application of the renders is as important as choosing the right materials for the rehabilitation works (Teutonico et al. 1994). Therefore, an appropriate rehabilitation requires the use of suitable materials and also good masons.

9.5 Conclusions

The use of more durable materials means buildings with a longer service life and thus a lower resources consumption. Reinforced concrete is a paramount structural material but prone to frequent early deterioration phenomena, like steel rebar corrosion and alkali-silica reaction. It is therefore essential to understand which specific measures must be taken at the design stage to obtain concrete structures with improved durability. On the other hand it is also important to understand how one can rehabilitate the existing concrete structures to extend their service life. Regarding the waterproofing of concrete with epoxy resins, it is important to remember the considerations made in Chap. 2 about the toxicity of these materials. The incorporation of wastes in concrete, addressed in Chap. 5 raises some doubts related to their durability and with regard to concretes containing C&DW aggregate contaminated by ASR, these doubts are more than justified. The renders and plasters in ancient buildings face serious problems regarding their rehabilitation, especially when it comes to substrates containing high amounts of soluble salts. In many cases some rehabilitation solutions can even worsen the problem. This chapter presented a series of procedures to identify the constituents of such renders in ancient buildings, as well as the requirements for their rehabilitation.

References

Adriano P, Silva A, Veiga R, Candeias A, Mirão J (2007) Determination of the composition of ancient mortars. 2° National Congress of Mortars, Lisbon

Albert LB (1995) Ten books on architecture. Oxford University Press, London

Arioglu N, Acun S (2006) A research about a method for restoration of traditional lime mortars and plasters: a staging system approach. Build Environ 41:1223–1230. doi:10.1016/j.buildenv.2005.05.015

Austin S, Robins P, Pan Y (1999) Shear bond testing of concrete repairs. Cem Concr Res 29:1067–1076. doi:10.1016/S0008-8846(99)00088-5

Bai J (2009) Durability of sustainable concrete materials. In: Khatib J (ed) Sustainability of construction materials. Woodhead Publishing Limited, Cambridge, pp 239–253

Bailey KL (1932) The Elder Pliny's chapter's on chemical subjects. Edward Arnold and Co, London

Bijen J (2000) Durability of engineering structures: design, repair and maintenance. Woodhead Publishing Limited, Cambridge

Carbo M, Reig F, Adelantado J, Martinez V (1996) Fourier transformed infrared spectroscopy and the analytical study of works of art for purposes of diagnosis and conservation. Anal Chim Acta 330:207–215. doi:10.1016/0003-2670(96)00177-8

Cechova E, Papayianni I, Stefanodou M (2008) The influence of lindseed oil on the properties of lime-based mortars. International conference HMC 08–hystorical mortars conference: characterization, diagnosis, repair and compatibility. LNEC, Lisbon

Charles-Gibergues A, Cyr M, Moisson M, Ringot E (2008) A simple way to mitigate alkali-silica reaction. Mater Struct 41:73–83. doi:10.1617/s11527-006-9220-y

Chechova E; Papayianni I; Stefanidou M (2010) Properties of lime-based restoration mortars modified by the addition of linseed oil. In: Valek J, Groot C, Hughes J (eds) Proceedings of the 2nd conference and of the final workshop of RILEM TC 203–RHM. RILEM Publications, Prague, pp 937–944

Corinaldesi V, Moriconi G, Tittarelli F (2003) Thaumasite: evidence for incorrect intervention in masonry restoration. Cem Concr Compos 25:1157–1160. doi:10.1016/S0958-9465(03) 00158-6

Cotrim H, Veiga R, Brito J (2008) Freixo palace: rehabilitation of decorative gipsum plasters. Constr Build Mater 22:41–49. doi:10.1016/j.conbuildmat.2006.05.060

Da Silva T (2007) Technical and economical analysis of some solutions for the retrofitting concrete structures on shore. Master thesis, IST/UTL, Lisbon

De Muynck W, Debrouwer D, De Belie N, Verstraete W (2008a) Bacterial carbonate precipitation improves the durability of cementitious materials. Cem Concr Resear 38: 1005–1014. doi:10.1016/j.cemconres.2008.03.005

De Muynck W, Cox K, De Belie N, Verstraete W (2008b) Bacterial carbonate precipitation as an alternative surface treatment for concrete. Constr Build Mater 22:875–885. doi:10.1016/ j.conbuildmat.2006.12.011

Elpida-Chrissy A, Eleni-Eva T, Elizabeth V (2008) Lime-pozzolan-cement compositions for the repair and strengthening of historic structures. International conference HMC 08–hystorical mortars conference: characterization, diagnosis, repair and compatibility, LNEC, Lisbon

Elsen J, Van Balen K, Mertens G (2010) Hydraulicity in historic lime mortars: a review. In: Valek J, Groot C, Hughes J (eds) Proceedings of the 2nd Conference and of the final workshop of RILEM TC 203–RHM. RILEM Publications, Prague, pp 129–145

Ferreira RM (2009) Service-life design of concrete structures in marine environments: a probabilistic based approach. VDM Verlag Dr. Muller Aktiengesellschaft & Co. KG, Germany

Flores-Colen I, De Brito J, Branco F (2007) On site evaluation of the adhesion strength for render materials. 2nd National Congress of Mortars, APFAC, Lisbon

Genestar C, Pons C (2003) Ancient covering plaster mortars from several convents and Islamic and Gothic places in Palma de Mallorca (Spain). Analytical characterization. J Cult Herit 291–298. doi: 10.1016/j.culher.2003.02.001

Gonçalves T (2007) Salt crystallization in plastered or rendered walls. PhD thesis, IST/UTL, Lisbon

Gonçalves T, Rodrigues J (2010) Renders for salt laden substrate masonry. 3rd National Congress of Mortars, APFAC, Lisbon

Gonçalves T, Pel L, Rodrigues J (2008) Worsening of dampness and salt damage after restoration: use of water repellent additives in plasters and renders. In: Proceedings of HMC 08 1st historical conference, LNEC, Lisbon

Grattan-Bellew P, Mitchell L, Margeson J, Min D (2010) Is alkali-carbonate reaction just a variant of alkali silica reaction ACR=ASR? Cem Concr Res 40:556–562. doi:10.1016/j.cemconres.2009.09.002

Groot C, Van Hees R, Wijffels T (2009) Selection of plasters and renders for salt laden masonry substrates. Constr Build Mater 23:1743–1750. doi:10.1016/j.conbuildmat.2008.09.013

Hill K (2000) Fats and oils as oleochemical raw materials. Pure Appl Chem 72:1255–1264. http://www.surfactantspectator.com/.../Harlheinx%20Hill%20on%20Oleochemicals.pdf

Hobbs DW (1988) Alkali-silica reaction in concrete. Thomas Telford, London

Ichikawa T (2009) Alkali-silica reaction, pessimum effects and pozzolanic effect. Concr Res 39:716–726. doi:10.1016/j.cemconres.2009.06.004

Jensen V (1993) Alkali aggregate reaction in Southern Norway. PhD thesis, Norwegian University of Science and Technology, Trondheim, Norway

Jonkers H, Thijssen A, Muyzer G, Copuroglu O, Schalangen E (2010) Application of bacteria as a self-healing agent for the development of sustainable concrete. Ecol Eng 36:230–235. doi:10.1016/j.ecoleng.2008.12.036

Justnes H, Ostnor T, Vila N (2004) Vegetable oils as water repellents for mortars. In: Proceedings of the 1st international conference of Asian Concrete Federation, Thailand, pp 689–698

Lambert P (2009) Sustainability of metals and alloys in construction. In: Khatib J (ed) Sustainability of Construction Materials. Woodhead Publishing Limited, Cambridge, pp 148–170

Lubelli B, Hees R, Groot C (2006a) Sodium chloride crystallization in a "salt transporting" restoration plaster. Cem Concr Res 36:1467–1474. doi:10.1016/j.cemconres.2006.03.027

Lubelli B, Hees R, Groot C (2006b) Investigation on the behavior of a restoration plaster applied on heavy salt loaded masonry. Constr Build Mater 20:691–699. doi:10.1016/j.conbuildmat.2005.02.010

Marques S (2005) Study about mortars for ancient buildings. Master thesis, University of Aveiro

Marques S, Ribeiro R, Silva L, Ferreira V, Labrincha J (2006) Study of rehabilitation mortars: Construction of a knowledge correlation matrix. Cem Concr Res 36:1894–1902. doi:10.1016/j.cemconres.2006.06.005

Medeiros M, Helene P (2008) Efficacy of surface hydrophobic agents in reducing water and chloride ion penetration in concrete. Mater Struct 41:59–71. doi:10.1617/s11527-006-9218-5

Moisson M (2005) Contribution á la maîtrise de la réaction alcali-silice par ajout de fines de granulats réactifs dans le béton. PhD thesis, INSA, France

Momayez A, Ehsani M, Ramezanianpour A, Rajaie H (2005) Comparison of methods for evaluating bond strength between concrete substrate and repair materials. Cem Concr Res 35:748–757. doi:10.1016/j.cemconres.2004.05.027

Mora E (2007) Life cycle, sustainability and the transcendent quality of building materials. Buil Environ 42:1329–1334. doi:10.1016/j.buildenv.2005.11.004

Moreira P (2006) Using polymeric coatings to improve the durability of concrete exposed to aggressive media. Master thesis, University of Minho

Morgan D (1996) Compatibility of concrete repair materials and systems. Constr Build Mater 10:57–67. doi:10.1016/0950-0618(95)00060-7

Pacheco-Torgal F, Jalali S (2009a) Pathology and rehabilitation of ancient gypsum plasters. Art & Constr J 223:24–26

Pacheco-Torgal F, Jalali S (2009b) Sulphuric acid resistance of plain, polymer modified, and fly ash cement concretes. Constr Build Mater 23:3485–3491. doi:10.1016/j.conbuildmat.2009.08.001

Pacheco-Torgal F, Gomes J, Jalali S (2008) Adhesion characterization of tungsten mine waste geopolymeric binder. Influence of OPC concrete substrate surface treatment. Constr Build Mater 22:154–161. doi:10.1016/j.conbuildmat.2006.10.005

Palladio A (1570) The four books on the architecture. Venice, Italy

Pedeferri P (1996) Cathodic protection and cathodic prevention. Constr Build Mater 10:391–402. doi:10.1016/0950-0618(95)00017-8

Poitvin P (1999) Limestone aggregates concrete, usefulness and durability. Cem Concr Res 21:89–97. doi:10.1016/S0958-9465(98)00047-X

Reis M, Silva A (1997) Alkali aggregate reaction. Recomendations to prevent concrete deterioration, LNEC ITM C23, Lisbon

Rowland PB (2006) Essays on Hormuz. http://www.dataxinfo.com/hormuz/essays/3.6.pdf

Sá AFG (2002) Renders based on air-lime mortars with vegetable fat. Monography N° 7, Construlink, IST/UTL, Lisbon

Sá AFG (2005) Renders for stone masonry. Master thesis, IST/UTL, Lisbon

Santos Silva A (2005) Concrete deterioration by ASR. Using fly ash to its prevention. PhD thesis, University of Minho/LNEC

Sarja A, Vesikar E (1996) Durability design of concrete structures. RILEM Report 14E & FN SPON. CRC Press, Otawa

Sims I, Brown B (1998) Concrete aggregates. In: Hewlett PC (ed) Lea's chemistry of cement and concrete, 4th edn. Arnold Publisher, London, pp 903–989

Sousa S, Santos Silva S, Velosa A, Rocha F (2010) Using tungsten mine waste to prevent ASR and DEF. Congress for innovation in sustainable construction, CINCOS'10, Centro Habitat, Cúria, Portugal, pp 217–228

Stanton TE (1940) Influence of cement and aggregate on concrete expansion. Eng News Record 1:50–61

Stolz H (2007) Oleochemicals—important additives for building protection. 2nd national congress of mortars, Lisbon

Teutonico J, Naccaig I, Burns C, Ashurst J (1994) The Smeaton project: factors affecting the properties of lime-based mortars. Bull Assoc Preserv Technol 25:32–49

Tittelboom K, De Belie N, De Muynck W, Verstraete W (2010) Use of bacteria to repair cracks in concrete. Cem Concr Res 40:157–166. doi:10.1016/j.cemconres.2009.08.025

Ugurlu E, Boke H (2008) The use of brick-lime plasters and their relevance to climatic conditions of historic bath buildings. Constr Build Mater 23:2442–2450. doi:10.1016/j.conbuildmat.2008.10.005

Veiga R (2003) Conservation mortars. In Proceedings of the 1st Meeting of Civil Engineering in the University of Aveiro, COM 103, LNEC, Lisbon

Vikan H, Justnes H (2006) Influence of vegetable oils on durability and pore structure of mortars. In: Proceedings of the seventh CANMET/ACI international conference on durability of concrete, Canada, pp 417–430

Wiering P (1984) Long time studies on the carbonation of concrete under normal outdoor exposure. RILEM symposium on durability of concrete under normal outdoor exposure, Hannover

Wood JGM, Johnson RA (1993) The appraisal and maintenance of structures with alkali-silica reaction. Struct Eng 71(2): 19–23

Zhang Z, Yao X, Zhu H (2010) Potential applications of geopolymers as protection coatings for marine concrete I. Basic properties. Appl Clay Sci 49:1–6. doi:10.1016/j.clay.2010.01.014

Chapter 10
Nanotechnology Achievements

10.1 General

The speech of the physicist Richard P. Feynmam, entitled "There's plenty of room at the bottom" (Feynman 1960), that took place in a meeting of the American Physical Society in 1959 at CalTech is considered to be the beginning of the nanotechnology era. This last term was nonetheless presented in 1974 by Professor Norio Taguchi, meaning the processing of materials, atom-by-atom or molecule-by-molecule (Taniguchi 1974). A more accurate definition of nanotechnology was presented in 1981 by Drexler (1981), such as the production with dimensions and precision between 0.1 and 100 nm. In medium terms nanotechnology involves the study at microcospic scale (1 nm = 1×10^{-9} m). As a comparison, one must realize that a human hair has 80,000 nm thickness and that the DNA double helix has 2 nm diameter. Between 1997 and 2003 the investment in nanotechnology increased at 40% reaching 35000 million euro (Andersen and Molin 2007). Some estimates predict that products and services related to nanotechnology could reach 1,000.000 million euro/year beyond 2015 (NSF 2001). The report RILEM TC 197-NCM, "Nanotechnology in construction materials" (Zhu et al. 2004), is the first document that synthesis in a clear manner the potential of nanotechnology in terms of the development of construction and building materials, namely:

- The use of nanoparticles, carbon nanotubes and nanofibres to increase the strength and durability of cimentitious composites as well as for pollution reduction;
- Production of cheap corrosion-free steel;
- Production of thermal insulation materials with a performance 10 times higher than commercial current options;
- Production of coats and thin films with self-cleansing ability and self-color change to minimize energy consumption;
- Production of nanosensores and materiais with sensing ability and self-repairing ability.

DOI: 10.1007/978-0-85729-892-8_10,

One of the most promising areas in the field of nanotechnology and also mentioned in the RILEM TC 197-NCM report relates to the replication of natural systems. The continuous improvement of these systems carried out over millions of years led to the development of materials and "technologies" with exceptional performance and fully bio-degradable. For instance, the abalone shells are made with 0.2 mm thickness layers, and each one is made by a "mortar" 0.5 μm thickness of calcium carbonate crystals bound altogether with a protein. The final result is a composite material with a toughness that is 3000 times the toughness of the calcium carbonate crystals (Li et al. 2004; Meyers et al. 2009). Mussels and barnacles, can produce a natural glue as good as synthetic adhesives that allows them to maintain a high adhesion to submerged rocks (Hedlund et al. 2004; Khandeparker and Chandrashekhar 2007; Kamino 2010). The great advantage relates to the fact that synthetic adhesives are based on epoxy, melamine-urea-formaldehyde, phenol or organic solvents. These compounds are toxic and responsible for eczema and dermatitis, and even cancer. Another example of the high performance of natural systems comes from the spider silk which possesses a strength/mass ratio that exceeds the steel ratio (Porter and Vollrath 2007; Lee et al. 2009; Harrington et al. 2010). Even comparing the performance of carbon nanotubes, a high-technology product with spider silk, one realizes how natural systems are well-optimized. First discovered in Russia in 1952, carbon nanotubes were some years later re-discovered in Japan. These materials have a tensile strength 100 times over steel but as a downside side they are extremely costly (20 to 1000 euros/g) (Man 2006). Another biomimicry-related finding relates to the coral reef formation. Those natural systems use sea water to produce calcium, magnesium and carbonate to generate a carbonate crystal (aragonite). The coral reef formation is a very complex process, dependent on several factors, the main being photosynthesis (Barnes 1970; Holcomb 2010). Recently the enterprise Calera has announced the possibility of producing calcium carbonate using sea water and CO_2 (Mitchell 2009; Geyer et al. 2009). Unfortunately the scientific information available so far is insufficient to take that for granted. However, the information concerning the technological process were not disclosed, only that the process is already being used in an experimental facility near the Moss Landing power plant (California), using carbon dioxide generated by the power plant to produce "cement". One high-impact application of nanotechnology in the field of energy consumption relates to the development of nanomaterials with very high-insulation performance, such as aerogel (Fig. 10.1) previously mentioned in Chap. 3.

10.2 Cementitious Composites with Enhanced Strength and Durability

10.2.1 Investigation of Portland Cement Hydration Products

Concrete is the most used construction material on Planet Earth and presents a higher permeability that allows water and other aggressive elements to enter, leading to carbonation and chloride ion attack, resulting in steel corrosion

Fig. 10.1 Aerogel

problems. Therefore, the nanoscale study of the hydration products (C–S–H, calcium hydroxide, ettringite, monosulfate, unhydrated particles and air voids), as a form to overcome durability issues, is a crucial step in the concrete eco-efficiency. Investigations in this field have already been carried out in recent years (Porro and Dolado 2005; Balaguru and Chong 2006). Mojumdar and Raki (2006) have already analyzed calcium silicate nanophase composites which will allow the future development of anti-corrosion and fire-retardant coatings. Until very recently electronic microscopy has allowed the understanding of the morphology, as well as the composition of hydration products. However, the use of nanotechnology currently allows the possibility of the knowledge of the elastic modulus by nanoindentation techniques (AFM). In nanoindentation a material with known characteristics is used to make a mark in another material with unknown properties and through the specific nature of this mark it is possible to infer the properties of the marked material. Recently Mondal (2008) used nanoindentation in cementitious phases and obtained the following elastic modulus: 35 MPa for the $Ca(OH)_2$ phase; 26 and 16 MPa for high- and low-stiffness C–S–H and 10 MPa for the porous phase. Other authors (Constantinides et al. 2003; Dejong and Ulm 2007; Constantinides and Ulm 2007) have already confirmed the existence of different types of CSH, low density, high density and ultra-high density. More recently some authors (Pellenq et al. 2009) from MIT have used nanotechnology to develop a molecular model for the hydration products of Portland cement. These authors confirm that the molecular model is in *excellent agreement* with experimental values obtained by nanoindentation techniques.

10.2.2 Composites with Nanoparticles

Nanoparticles have a high-surface area to volume ratio providing high-chemical reactivity. They act as nucleation centers, contributing to the development of the hydration of Portland cement. Most investigations use nanosilica while some already used nano-Fe_2O_3. The production of nanoparticles can be obtained either through a high-milling energy (Sobolev and Ferrada-Gutierrez 2005) or by chemical synthesis (Lee and Kriven 2005). Porro et al. (2005) mentioned that the use of nanosilica particles increases the compression strength of cement pastes. The same authors state that the phenomenon is not due to the pozzolanic reaction, because calcium hydroxide consumption was very low but, instead, due to the increase use of silica compounds that contributes to a denser microstructure. According to Lin et al. (2008), the use of nanosilica on sludge/fly ash mortars, compensates the negative effects associated to the sludge incorporation in terms of setting time and initial strength. Sobolev et al. (2008) reported that nanosilica addition led to an increase of strength by 15% to 20%. Other authors (Gaitero 2008; Gaitero et al. 2009) believe that nanosilica leads to an increase of C–S–H chain dimension and also to an increase of C–S–H stiffness. Chen and Lin (2009) used nanosilica particles to improve the performance of sludge/clay mixtures for tile production. The results show that nanoparticles improved the reduction of water absorption and led to an increase of abrasion and impact strength. Others (Vera-Agullo et al. 2009) also confirm that the use of nanoparticles (nanotubes, nanofibers, nanosilica or nanoclay) is responsible for a higher hydration degree of cementitious compounds, as long as a higher nanoparticle dispersion can be achieved. Nasibulin et al. (2009) reported an increase in strength by 2 to 40 times for electric conductivity, which means a high potential for sensing ability. Several authors confirm the suitability of mortars with Fe_2O_3 nanoparticles to act as sensing materials (Li et al. 2004; Qing et al. 2008; Lin et al. 2008). Chaipanich et al. (2010) mentioned that 1% of carbon nanofibres (by binder mass) can compensate the strength reduction associated with the replacement of 20% fly ash. Gdoutos–Konsta et al. (2010) also studied the effect of carbon nanofibres on cement pastes (0.08% by binder mass) observing an increase in the mechanical strength. Those authors used ultra-sounds to achieve a high-nanofibre dispersion stating that this is a crucial step in order to obtain a high performance of nanotubes in the cement matrixes. Nevertheless, the fact that carbon nanotubes are not cost-efficient prevents the increase of its use in commercial applications in a near future.

10.3 Photocatalytic Applications

The most known application of nanomaterials in the construction industry relates to the photocatalytic capacity of semiconductor materials. Several semiconductors materials, such as TiO_2, ZnO, Fe_2O_3, WO_3 and CdSe, possess photocatalytic

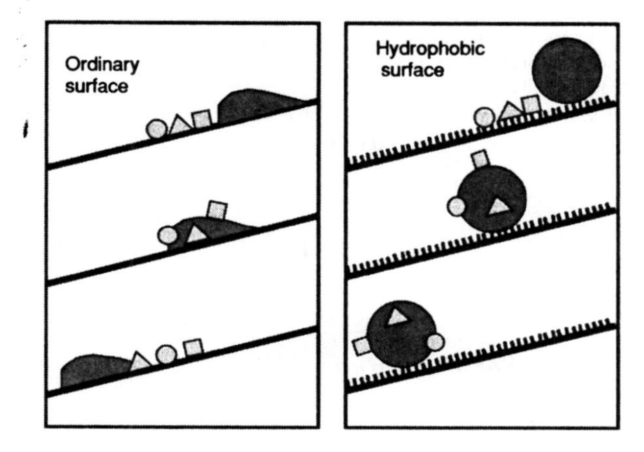

Fig. 10.2 Lotus effect (Benedix et al. 2000)

capacity (Makowski and Wardas 2001). However, TiO_2 is the most used of all because of its low toxicity and stability (Djebbar and Sehili, 1998). Titanium dioxide can crystallize as rutile, anatase and brookite, being the first form the most stable (thermodynamically speaking), it is also the most available form (it is the 9th most abundant element in the Earth crust), being currently used as additive in the painting industry.

The anatase and brookite forms are meta-stable and can be transformed into rutile by thermal treatment. Being a semiconductor with photocatalytic capacity, when TiO_2 is submitted to UV rays (320–400 nm), in the presence of water molecules (Husken et al. 2009), it leads to the formation of hydroxyl radicals (OH) and superoxide ions (O_2^-). Those highly-oxidative compounds react with dirt and inorganic substances promoting their disintegration. The photocatalysis of TiO_2 is also responsible for the reduction of the contact angle between water droplets and a given surface, leading to super-hydrofobic or super-hydrophilic surfaces increasing their self-cleansing capacity. Water repellent surfaces are one of the features of natural systems as it happens in the leaves of the lotus plant, whose microstruture allows self-cleansing ability (Fig. 10.2). According to Fujishima et al. (2008), the potential of photocatalysis can be perceived by the high number of citations (almost 3700) of a related paper published on *Nature* in 1972, as well as by the number of papers concerning photocatalysis investigations that increased in an exponential pattern between 1997 and 2007. Another form to evaluate the potential of this technology is by knowing that the Japanese Corporation TOTO Ltd has already issued 1200 international patent requests in this field. So far 500 have been approved. The applications related to photacatalysis cover five different groups (Fig. 10.3).

Considering the cost to clean graffiti paintings (in Los Angeles city this could amount to 100 million euro/year (Castano and Rodriguez 2003))one can realize the huge potential of the photocatalytic capacity of nanomaterials.

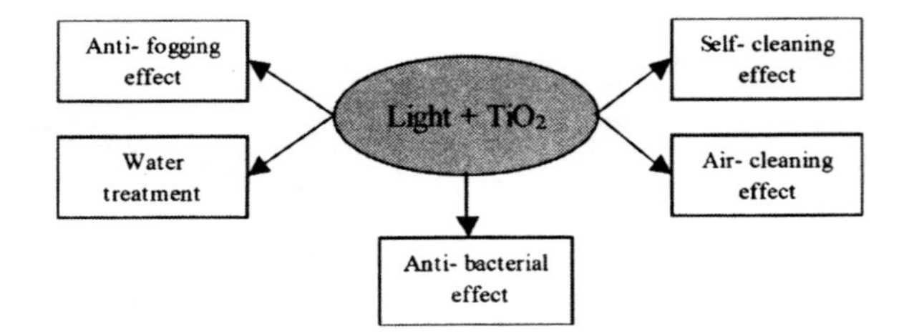

Fig. 10.3 Major areas of activity in titanium dioxide photocatalysis (Benedix et al. 2000)

10.3.1 Self-Cleaning Ability

Although self-cleaning properties of photocatalysts materials are known since the 60s (Fujishima and Honda 1972), only recently they start to be used in a wide-scale (Fujishima et al. 1999). Cassar and Pepe (1997) patented a concrete block with self-clening ability (Table 10.1).

The first application of self-cleaning concrete took place in the church "Dives in Misericordia" in Rome (Fig. 10.4). This building was designed by the Arq° Richard Meyer and officially opened in 2003. It is composed of 346 pre-stressed concrete blocks made with white cement and TiO_2 (binder 380 kg/m^3 and W/B = 0.38) (Cassar et al. 2003).

Visual observations carried out six years after its construction revealed only slight differences between the white color of the outside concrete surfaces and the inside block surfaces (Chen and Poon 2009b). Diamanti et al. (2008) studied mortars containing TiO_2 having noticing reductions in the contact angle between water and solid surface of almost 80%. Ruot et al. (2009) mentioned that the photocatalytic activity is dependent on the matrix properties. Increasing the TiO_2 content in cement pastes above 1% leads to a proportional increase in the photocatalytic activity, as for mortars a TiO_2 content increase just lead to a very small increase in the photocatalytic activity (Fig. 10.5). Those authors suggest that most TiO_2 particles in mortars are not reached by UV radiation.

The use of TiO_2 thin films on tiles or glasses has significant potential in terms of self-cleaning ability. According to Fujishima et al. (2008) Japan buildings must be cleaned at least every five years to maintain a good appearance, while that covered with self-cleaning tiles should remain clean over a span of 20 years without any maintenance.

10.3.2 Air Pollution Reduction

The subject of air pollutants such as VOCs released from building materials have already been addressed in . Chap. 2. In the last years several investigations have been carried out in order to use photocatalysis to reduce air pollution. The reaction

Table 10.1 Patent for paving tile comprising an hydraulic binder and photocatalyst particles (Cassar and Pepe 1997)

Field of the invention	Hydraulic binder, dry premix, cement composition having improved property to maintain the brilliance and color quantity and to prevent aesthetic degradation
Working principle/ product requirements	Use of photocatalyst particles able to oxidize air and environmental pollutants;
	Use of a photocatalyst which is able to oxidize in the presence of light air and environmental polluting substances for the preparation of an hydraulic binder for manufacturing paving tiles that maintain after installation for a longer time brilliance and color quantity;
	Use of a dry premix containing a hydraulic binder and a photocatalyst that is able to oxidize in the presence of light air and environmental polluting substances for manufacturing paving tiles that maintain after installation for a longer time brilliance and color quantity
Binder	Hydraulic binder
	Cement (white, gray or pigmented)
	Cement used for debris dams
	Hydraulic lime
Photocatalyst	TiO_2 without further requirements-TiO_2 or a precursor thereof, mainly in the form of anatase
	TiO_2 with anatase structure for at least 25, 50 and 70%
	Blend of anatase and rutile TiO_2 having a ratio 70:30
	TiO_2 doped with one or more atoms different from Ti
	TiO_2 doped with one or more atoms selected from Fe(III), Mo(V), Ru(III)
	Os(III), Re(V), V(V), Rh(III)
	Photocatalyst selected from the group consisting of tungstic oxide (WO_3), strontium titanate ($SrTiO_3$) and calcium titanate ($CaTiO_3$)
Amount of photocatalyst	0.01–10% by weight
	0.1% by weight with respect to the binder
	0.5% by weight with respect to the binder

Fig. 10.4 Church "Dives in Misericordia", Rome

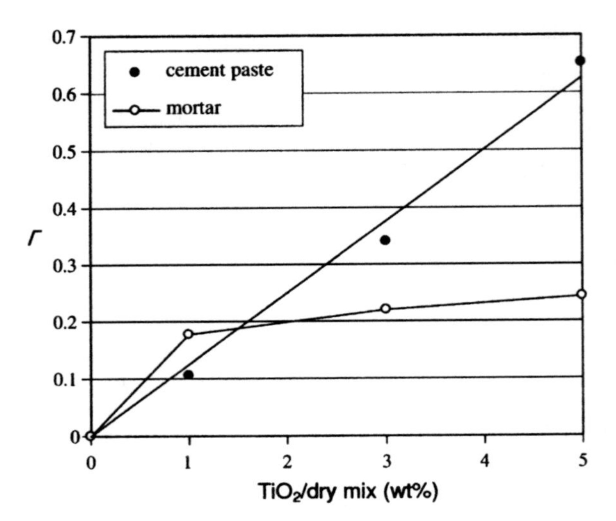

Fig. 10.5 Photocatalytic activity of cement-based materials versus the TiO_2 content, within the first 7 h of illumination (Ruot et al. 2009)

of photocatalytic oxidation of pollutants, generates water and carbon dioxide as by-products. Murata et al. (1997) patented a paving block for the reduction of air pollution (Table 10.2):

Zhao and Yang (2003) mentioned a high-photocatalytic capacity for indoor air pollution reduction when using P25 TiO_2 (70% anatase + 30% rutile) with 300 nm diameter and a specific surface of 50 m^2/g. Yu (2003) studied cementitious paving blocks for NO_x reduction, noticing that the photocatalytic capacity is reduced by the presence of dust, grease or plastic gum, thus suggesting that these blocks should not be placed in pedestrian areas. Maier et al. (2005) reported a fast pollution reduction in indoor air by the use of gypsum plasters containing 10% TiO_2 (Fig. 10.6). Those authors mentioned that although air pollution reduction is dependent on the UV intensity, nevertheless, visible light still allows acceptable degradation rates. Those plasters were used to cover some bedrooms in Sweden, being responsible for a reduction on VOC of about 1/3 (to 26 $\mu g/m^3$).

Strini et al. (2005) mentioned that TiO_2 thin films have a photocatalytic capacity which is 3 to 10 times higher than for TiO_2 based cementitious composites. In 2006 the results of the PICADA (2006) project "Photo-catalytic innovative coverings applications for de-pollution assessment" aiming to the development of TiO_2-based coatings for self-cleaning and air pollution reduction coatings were disclosed. This consortium gathered 8 partners (Italcementi, Millenium Chemicals, AUT, NCSRD, CNR ITC, CSTB, Dansk Beton Teknik and GMT) and was financed in 2.3 million euro by the EU (Gurol, 2006). Aside from the study of small specimens in laboratory the PICADA project also covered pilot tests at macro-scale (1:5) in order to reply the effect of a street by using (18×5.18 m^2) "walls" and an artificial NO_x pollution source (Fig. 10.7). The results showed a reduction in the NO_x emissions between 40% to 80%. However, results published in a scientific journal mentioned NO_x reductions between 36.7%

Table 10.2 Patent for paving block capable of reducing NO_x (Murata et al. 1997)

Name of the patent	NO_x-cleaning paving block
Field of the invention	NO_x-cleaning paving block with enhanced NO_x-cleaning capability due to an increased efficiency of fixing NO_x from the air and increased pluvial NO_x-cleaning efficiency and is provided with a non-slip property, wear resistance and decorative property
Working principle/product requirements	NO_x-cleaning paving block comprising a surface layer which contains TiO_2 and a concrete made base layer
	NO_x-cleaning paving block with or without adsorbing material in the surface layer
	Replacement of the sand used by 10–50% of glass grains or silica sand having a particle size of 1–6 mm
	Surface layer having a void fraction of 10–40% and water permeability of 0.01 cm/s
	NO_x-cleaning paving block roughened with a surface roughening tool
Binder	Cement
Photocatalyst	TiO_2 without further requirements
Amount of photocatalyst	0.6–20% by weight
	5–50% by weight with respect to the binder
Adsorbing materials	Zeolite, magadiite, petalite and clay
Thickness of the surface layer	2–15 mm

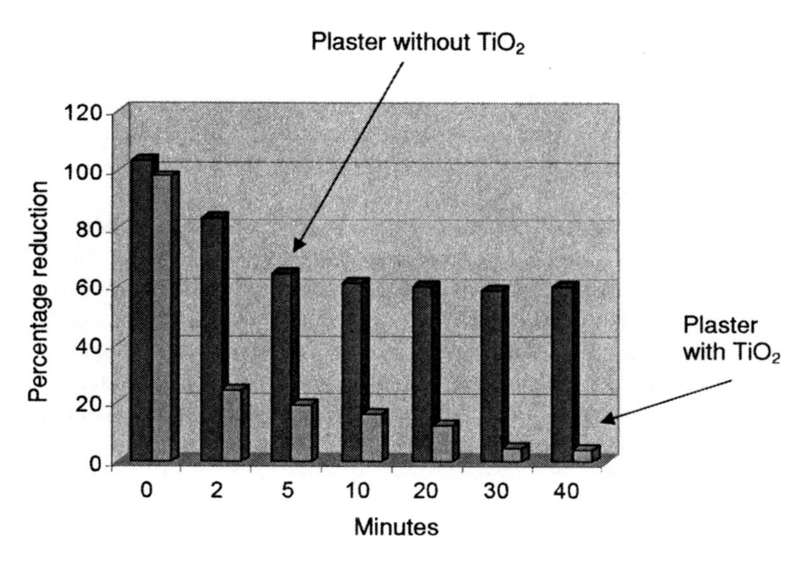

Fig. 10.6 Degradation of formaldehyde in plasters containing TiO_2 (Maier et al. 2005)

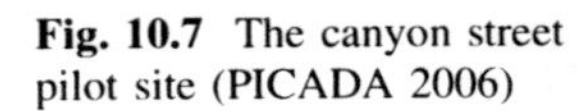

Fig. 10.7 The canyon street pilot site (PICADA 2006)

to 42% (Maggos et al. 2008). The use of a three-dimensional numerical model (MIMO) based on the data generated in the pilot test allowed the insertion of win velocity and temperature in order to predict the reductions in air pollution by the photocatalytic activity of the facade coatings containing TiO_2.

In another macro-scale test carried out in the PICADA project, the ceiling of an underground car park (322 m^2) was painted with TiO_2-based paint. Then the park was sealed and polluted by the exhaust gas from a single car. The results showed a 20% reduction on NO_x emissions, due to the photocatalytic capacity of the paint in the ceiling. Wang et al. (2007) confirmed that in the last few years a lot of investigations has been made about the reduction of indoor air pollution when using UV radiation but very few have analyzed the possibility of using photocatalysts active under visible light. Poon and Cheung (2007) mentioned that TiO_2 cementitious composites with increased porosity show a high NO_x emissions reduction. Those authors compared the performance of several TiO_2 forms, concluding that although P25 is much more reactive it does not have a very high performance/cost ratio. Guerrini and Peccati (2007) mentioned a case of a street in Bergamo, Italy, paved with blocks (12,000 m^2) of photocatalytic properties where high reductions of NO_x emissions (45%) have been reported. In Antwerp a park with 10,000 m^2 of semiconductors paving blocks also showed a reduction in NO_x emissions (Beeldens 2007). In Tokyo cement mixtures containing TiO_2 colloidal solutions were used to coated several road areas (Fig. 10.8). The results obtained in an area of 300 m^2 show a 50 mg to 60 mg/day NO emissions degradation (Fujishima et al. 2008).

Auvinen and Wirtanen (2008) studied the reduction of VOCs in indoor air when using paints with TiO_2 applied in several substrates (glass, gypsum and polymer) noticing that the substrate does not influences the photocatalytic reaction. Those authors mentioned that organic additives must not be used for this paints because they will be damaged by the radical hydroxyls. Also that the photocatalytic reactions generates not only water and CO_2 but also other pollutants that are harmful for human health. Other authors (Demeestere et al. 2008) confirm the reductions of NO_x emissions between 23% and 63% when using TiO_2-based tiles. They also reported that the accumulation of reaction products generated in the oxidation process reduces the photocatalytic activity. For the production of TiO_2-based cementitious composites some authors (Husken et al. 2009) recommend the use of TiO_2 as a solution with the mixing water because it allows for a better

Fig. 10.8 Using TiO_2 photocatalytic material on roadway for pollution reduction (Fujishima et al. 2008)

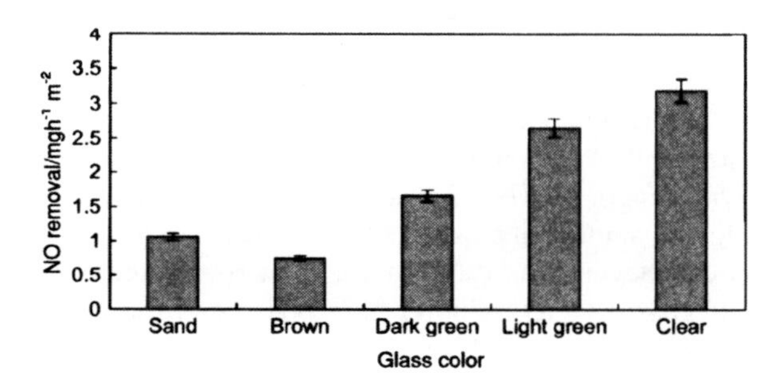

Fig. 10.9 Comparison of NO removal by samples of different glass colors based on 7-day curing age testing. (Chen and Poon 2009a)

dispersion than if it is mixed with the cement. Also that a semiconductor with a high-specific surface gives better results than the use of a superior volume of low-specific surface semiconductor. These authors used a specific surface TiO_2 (between 0.7 m^2/g to 1.5 m^2/g) which is much lower than the P25 form. Chen and Poon (2009b) found out that when using grounded glass for partial sand replacement the photocatalytic capacity, increases as much as 3 times for lighter glass (Fig. 10.9). Those authors suggest that glass particles can allow light to enter more deeply in the mortar leading to a higher oxidation rate.

Kolarik et al. (2010) confirmed that the photocatalytic reaction is a good way to reduce VOCs in indoor air. Ramirez et al. (2010) mentioned that the substrate porosity influences the photocatalytic reaction when TiO_2 thin films are being used and that a high-porosity surface leads to a high photocatalytic reaction. Ballari et al. (2010) presented a model that can predict NO_x emissions reduction using concrete with TiO_2 particles. Hassan et al. (2010) used 41 MPa concrete blocks covered with 1 cm TiO_2-based mortar layer in order to evaluate nanoparticles removal by abrasion tests and thus reducing the photocatalytic capacity in NO_x emissions reduction. Those authors mentioned that even after 20,000 abrasion cycles the NO_x emissions degradation and reduction remained stable.

10.3.3 Bactericidal Capacity

One of the most important applications of materials with photocatalytical properties concerns the destruction of fungi and bacteria. Indoor fungi and bacteria proliferation are one of the main causes responsible for construction materials degradation and also for health problems (Zyska 2001; Santucci et al. 2007; Wiszniewska et al. 2009; Bolashikov and Melikov 2009) because fungi are responsible for mycotoxins growth (Reboux et al. 2010). Saito et al. (1992) studied the addition of TiO_2 powder with an average size of 21 nm (30% rutile and 70% anatase) to a bacterial colony. The results showed that 60 to 120 min were sufficient to destroy all the bacteria. Those authors state that using bigger TiO_2 particles reduces the bactericidal capacity and that the best results are obtained for a TiO_2 concentration between 0.01 to 10 mg/ml. Huang et al. (2000) also confirmed that using lower dimension TiO_2 particles leads to a faster bacterial destruction. Those authors noticed that bacterial destruction begins after 20 min of UV radiation exposition, being that after 60 min all the bacteria have been destroyed. They also reported that after the destruction has been initiated the fact that UV radiation is stopped does not reduce the bactericidal effect (Fig. 10.10).

Some authors (Kuhn et al. 2003) believe that the bactericidal capacity associated with TiO_2 photocatalysis is dependent on the use of UV-A radiation with a wavelength between 320 and 400 nm, being that UV-C type is only effective if the light is applied in a direct manner, thus preventing the treatment of less illuminated areas. Seven et al. (2004) found that zinc-based photocatalysis is as bactericidal as effective as TiO_2. Cho et al. (2004) confirmed that hydroxyl radicals are mainly responsible for the bactericidal capacity of semiconductors photocatalysts. Those authors mentioned that hydroxyl radicals have a destruction capacity of *E.coli* bacteria which is 1000 to 10000 times more effective than the chemical disinfection products. Vohra et al. (2006) used silver-doped TiO_2 noticing a 100% bacteria destruction just after 2 min, this compares in a most favorable manner with current TiO_2 which took 2 h to achieve the same destruction level. The bactericidal capacity is reduced over time because of the accumulation of dead bacteria and viruses (Bolashikov and Melikov 2009). Other authors (Chen et al. 2009) used wood specimens coated with a TiO_2 thin film (1.5 mg/cm^2) noticing that the photocatalytic reaction prevented fungi growth. Calabria et al. (2010) analyzed the application of TiO_2 thin films (20 to 50 nm thickness) by the sol–gel process in adobe blocks as a way to increase their water absorption and the bactericidal capacity. Those authors mentioned that TiO_2 thin films could be more cost-effective than current commercial paints. One of the main disadvantages of the bactericidal effect associated with photocatalysis relates to the need of UV radiation with a wavelength between 200 to 400 nm, however, recent findings show some possibilities in the development of composite materials with photocatalytic properties even when exposed to visible light (Dunnill et al. 2009; Chen et al. 2010). The use of titanium and trioxide tungsten-based films showed high-photocatalytic capacity under visible light above 400 nm (Song et al. 2006;

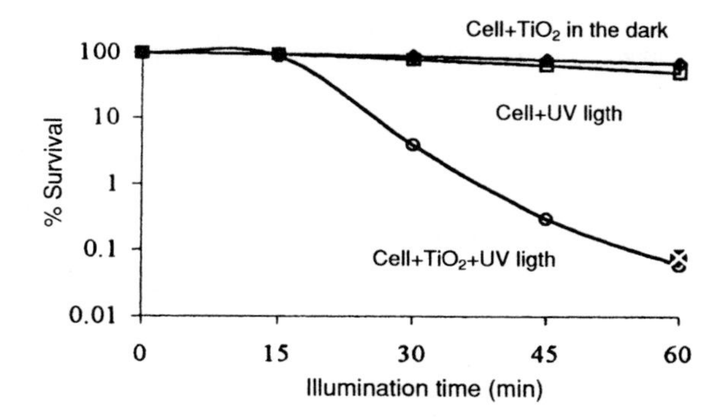

Fig. 10.10 The effect of TiO_2 photocatalytic reaction on cell viability. (Huang et al. 2000)

Saepurahman and Chong 2010). Herrmann et al. (2007) mentioned several questions that should be addressed in a near future:

- Use of semiconductors which are not TiO_2 based;
- Photocatalysis activation using visible light;
- Development of semiconductors with improved bactericidal capacity.

10.4 Conclusions

Nanotechnology has the potential to be the key to a brand new world in the field of construction and building materials. Although the replication of natural systems is one of the most promising areas of this technology, scientists are still trying to grasp their astonishing complexities. Recent years showed an intensive use of the potential of some photocatalytic nanomaterials, by the development of products with self-cleaning ability, products capable of reducing air pollution and with bactericidal capacity. The results of these investigations show that the titanium dioxide is the most widely used semiconductor in the photocatalytic reaction due to its low toxicity and stability. They also show that the efficiency of the photocatalytic reaction is dependent on the type of TiO_2, being that a mixture of rutile (30%) and anatase (70%) seems to be the most reactive. The use of TiO_2 with a high-specific surface area also shows a higher reactivity. As to the use of TiO_2 dispersed in cement matrix it is less effective than as thin films.

References

Andersen M, Molin M (2007) NanoByg- a survey of nanoinnovation in Danish construction. ISBN 978-87-550-3589-8, Rise National Laboratory, Technical University of Denmark

Auvinen J, Wirtanen L (2008) The influence of photocatalytic interior paints on indoor air quality. Atmosp Environ 42:4101–4112. doi:10.1016/j.atmosenv.2008.01.031

Balaguru P, Chong K (2006) Nanotechnology and concrete: Research opportunities. ACI FALL 2006 Convention, Nanotechnology of Concrete: Recent Developments and Future Perspectives, Code 76031.ACI

Ballari M, Hunger M, Husken G, Brouwers H (2010) NO_x photocatalytic degradation employing concrete pavement containing titanium dioxide. Appl Catal B: Environ 95:245–254. doi: 10.1016/j.apcatb.2010.01.002

Barnes D (1970) Coral skeletons: An explanation of their growth and structure. Science 170:1305–1308. doi:10.1126/science.170.3964.1305

Beeldens A (2007) Air purification by road materials: results of the test project. In: Baglione P, Cassar I (ed) RILEM Int. Symp. on photocatalysis environment and construction materials, 187–194, Italy

Benedix R, Dehn F, Quaas J, Orgass M (2000) Application of titanium dioxide photocatalysis to create self-cleaning building materials. Lacer 5:158–168

Bolashikov Z, Melikov A (2009) Methods for air cleaning and protection of building occupants from airborne pathogens. Build Environ 44:1378–1385. doi:10.1016/j.buildenv.2008.09.001

Calabria J, Vasconcelos W, Daniel D, Chater R, Mcphail D, Boccaccini A (2010) Synthesis of sol-gel titania bactericidal coatings on adobe brick. Constr Build Mater 24:384–389. doi: 10.1016/j.conbuildmat.2009.08.020

Cassar I, Pepe C (1997) Paving tile comprising an hydraulic binder and photocatalytic particles. EP-patent 1600430A1, Italcementi, Italy

Cassar L, Pepe C, Tognon G, Guerrini G, Amadelli R (2003) White cement for architectural concrete possessing photocatalytic properties. 11th International Congress on the Chemestry of Cement, Durban

Castano V, Rodriguez R (2003) A nanotechnology approach to high performance anti-graffiti coatings. Presentation at the Nanotechnology in Crime Prevention Conference, London

Chaipanich A, Nochaya T, Wongkeo W, Torkittikul P (2010) Compressive strength and microstructure of carbon nanotubes-fly ash cement composites. Mater Science Eng A527: 1063–1076. doi:10.1016/j.msea.2009.09.039

Chen L, Lin D (2009) Applications of sewage sludge ash and nano-SiO_2 to manufacture tile as construction material. Const Build Mater 23:3312–3320. doi:10.1016/j.conbuildmat.2009.06.049

Chen J, Poon C (2009a) Photocatalytic construction and building materials: from fundamentals to applications. Build Environ 44: 1899–1906. doi:10.1016/j.buildenv.2009.01.002

Chen J, Poon C (2009b) Photocatalytic activity of titanium dioxide modified concrete materials—Influence of utilizing recycled glass cullets as aggregates. J Environ Manag 90: 3436–3442. doi:10.1016/j.jenvman.2009.05.029

Chen F, Yang X, Wu Q (2009) Antifungal capability of TiO_2 coated film on moist wood. Build Environ 44:1088–1093. doi:10.1016/j.buildenv.2008.07.018

Chen F, Yang X, Mak H, Chan D (2010) Photocatalytic oxidation for antimicrobial control in built environment: A brief literature overview. Build Environ 45:1747–1754. doi:10.1016/j.buildenv.2010.01.024

Cho M, Chung H, Choi W, Yoon J (2004) Linear correlation between inactivation of E.coli and OH radical concentration in TiO_2 photocatalytic disinfection. Water Res 38: 1069–1077. doi: 10.1016/j.buildenv.2010.01.024

Constantinides G, Ulm F (2007) The nanogranular nature of C–S–H. J Mech Phys Sol 55:64–90. doi:10.1016/j.jmps.2006.06.003

Constantinides G, Ulm F, Vliet K (2003) On the use of nanoindentation for cementitious materials. Mater Struct 36:191–196. doi:10.1016/j.buildenv.2010.01.024

Dejong M, Ulm F (2007) The nanogranular behavior of C–S–H at elevated temperatures (up to 700°C). Cem Concr Res 37:1–12. doi:10.1016/j.cemconres.2006.09.006

Demeestere K, Dewulf J, De Witte B, Beeldens A, Van Langenhove H (2008) Heterogeneous photocatalytic removal of toluene from air on building materials enriched with TiO_2. Build Environ 43:406–414. doi:10.1016/j.buildenv.2007.01.016

Diamanti M, Ormellese M, Pedeferri M (2008) Characterization of photocatalytic and superhydrophilic properties of mortars containing titanium dioxide. Cem Concr Res 38: 1343–1353. doi:10.1016/j.cemconres.2008.07.003

Djebbar K, Sehili T (1998) Kinetics of Heterogeneous photocatalytic decomposition of 2, 4-Dichlorophenoxyacetic acid over TiO2 and ZnO in aqueous solution. Pest Sci 54:269–276. doi:10.1016/j.jcis.2005.08.007

Drexler K (1981) Molecular engineering: An approach to the development of general capabilities for molecular manipulation. Proc. Natl. Acad. Sci USA 78:5275–5278

Dunnill C, Ziken Z, Pratten J, Wilson M, Morgan D, Parkin I (2009) Enhanced photocatalytic activity under visible light in N-doped TiO2 thin film produced by APCVD preparations using t-butylamine as a nitrogen source and their potential for antibacterial films. J Photochem Photobiol A: Chem 207:244–253. doi:10.1016/j.jphotochem.2009.07.024

Feynman R (1960) There's plenty of room at the bottom (reprint from the speech given at the annual meeting of the West Coast section of the American Physical Society). Eng Sci 23:22–36

Fujishima A, Honda K (1972) Electrochemical photolysis of water at a semiconductor electrode. Nat 238:37–38. doi:10.1038/238037a0

Fujishima A, Hashimoto K, Watanabe T (1999) Photocatalysis. Fundamentals and its Applications, BCK Inc, Japan

Fujishima A, Zhang X, Tryk D (2008) TiO_2 photocatalys and related surface phenomena. Surface Sci Reports 63:515–582. doi:10.1016/j.surfrep.2008.10.001

Gaitero J (2008) Multi-scale study of the fibre matrix interface and calcium leaching in high performance concrete. PhD Thesis, Centre for Nanomaterials Applications in Construction of Labein-Tecnalia, Spain

Gaitero J, Zhu W, Campillo I (2009) Multi-scale study of calcium leaching in cement pastes with silica nanoparticles. Nanotechnology in Construction 3, Springer, Berlin, Heidelberg

Gdoutos-Konsta M, Metaxa Z, Shah S (2010) Highly dispersed carbon nanotube reinforced cement based materials. Cem Concr Res 40:1052–1059. doi:10.1016/j.cemconres.2010.02.015

Geyer R, Del Maestro C, Rohloff A (2009) Greenhouse gas footprint analysis of the Calera process. University of California, California

Guerrini G, Peccati E (2007) Photocatalytic cementitious roads for depollution. In: Baglione P, Cassar I (ed) RILEM Int. Symp. on photocatalysis environment and construction materials, 179–186, Italy

Gurol M (2006) Photo-catalytic construction materials and reduction in air pollutants. Sacramento State Center for California Studies, The California State University. http://www.csus.edu/calst/government_affairs/reports/PHOTO-CATALYTIC.pdf. Accessed 3 July 2011.

Harrington M, Masic A, Holten-Andersen N (2010) Iron-clad fibers: A metal-based biological strategy for hard flexible coatings. Sci 216–220. doi: 10.1126/science.1181044

Hassan M, Dylla H, Mohammad L, Rupnow T (2010) Evaluation of the durability of titanium dioxide photocatalyst coating for concrete pavement. Const Build Mater 24:1456–1461. doi: 10.1016/j.conbuildmat.2010.01.009

Hedlund J, Berglin M, Sellborn A, Andersson M, Delage L, Elwing H (2004) Marine adhesives as candidates for new biomaterial. Transactions—7th World Biomaterials Congress

Herrmann J, Duchamp C, Karkmaz M, Hoai B, Lachheb H, Puzenat E, Guillard C (2007) Environmental green chemistry as defined by photocatalysis. J Hazard Mater 146:624–629. doi:10.1016/j.jhazmat.2007.04.095

Holcomb M (2010) Coral calcification: insights from inorganic experiments and coral responses to environmental variables. PhD Thesis, Massachusetts Institute of Technology MIT, Cambridge

Huang Z, Maness P, Blakem D, Wolfrum E, Smolinski S, Jacoby W (2000) Bactericidal mode of titanium dioxide photocatalysis. J Photochem Photobiol A:Chem 130:163–170. doi:10.1016/S1010-6030(99)00205-1

Husken G, Hunger M, Brouwers H (2009) Experimental study of photocatalytic concrete products for air purification. Build Environ 44:2463–2474. doi:10.1016/j.buildenv.2009.04.010

Kamino K (2010) Molecular design of barnacle cement in comparison with those of mussel and tubeworm. J Adhesion 86:96–110

Khandeparker L, Chandrashekhar A (2007) Underwater adhesion: The barnacle way. Int J Adhesion Adhesives 27:165–172. doi:10.1016/j.ijadhadh.2006.03.004

Kolarik B, Wargocki P, Skorek-Osikowska A, Wisthaler A (2010) The effect of a photocatalytic air purifier on indoor air quality quantified using different measuring methods. Buil Environ 45:1434–1440. doi:10.1016/j.buildenv.2009.12.006

Kuhn K, Chaberny I, Massholder K, Stickler M, Benz V, Sootag H, Erdinger L (2003) Disinfection of surfaces by photocatalytic oxidation with titanium dioxide and UVA light. Chemosphere 53:71–77. doi:10.1016/S0045-6535(03)00362-X

Lee S, Kriven W (2005) Synthesis and hydration study of Portland cement components prepared by organic steric entrapment method. Mater Struct 38:87–92. doi:10.1007/BF02480579

Lee S, Pippel E, Gosele U, Dresbach C, Qin Y, Chandran C, Brauniger T, Hause G, Knez M (2009) Greatly increased toughness of infiltrated spider silk. Sci 324:488–492. doi:10.1126/science.1168162

Li H, Xiao H, Ou J (2004) A study on mechanical and pressure-sensitive properties of cement mortar with nanophase materials. Cem Concr Res 34: 435–438. doi:10.1016/j.cemconres.2003.08.025

Lin D, Lin K, Chang W, Luo H, Cai M (2008) Improvements of nano-SiO_2 on sludge/fly ash mortar. Waste Manag 28:1081–1087. doi:10.1016/j.wasman.2007.03.023

Maggos T, Plassais A, Bartzis J, Vasilakos C, Moussiopoulos N, Bonafous L (2008) Photocatalytic degradation of NO_X in a pilot street configuration using TiO_2-mortar panels. Environ Monitor Assess 136:35–44. doi:10.1007/s10661-007-9722-2

Maier W, Nilsson C, Holzer M, Lind J, Rosebom K (2005) Photocatalytic plaster for indoor air purification. 1st National Congress of Construction Mortars, APFAC, Lisbon

Makowski A, Wardas W (2001) Photocatalytic degradation of toxins secreted to water by cyanobacteria and unicellular algae and photocatalytic degradation of the cells of selected microorganisms. Current Topics in Biophysics 25: 19–25. www.nanoes.com.hk/.../Water/Photocatalytic_degration-of-toxins.pdf

Man S (2006) Nanotechnology and construction. Nanoforum report, Institute of Nanotechnology

Meyers M, Lim C, Nizam B, Tan E, Seki Y, Mckittrick J (2009) The role of organic intertile layer in abalone nacre. Mater Sci Eng C29:2398–2410. doi:10.1016/j.msec.2009.07.005

Mitchell S. (2009) Capturing carbon. Concrete Construction. World of Concrete 54:104

Mojumdar S, Raki L (2006) Synthesis, thermal and structural characterization of nanocomposites for potential applications in construction. J Therm Anal Calorimetry 86: 651–657. 10.1007/s10973-006-7720-1

Mondal P (2008) Nanomechanical properties of cementitious materials. PhD Thesis in Civil and Environment Engineering. Northwestern University, Illinois

Murata Y, Tawara H, Obata H, Murata K (1997) NO_x-cleaning paving block. European patent 0 786 283 A1. Mitsubhisi Materials Corporation, Japan

Nasibulin A, Shandakov S, Nasibulina L, Cwirzen A, Mudimela P, Habermehl-Cwirzen K, Grishin D, Gavrilov Y, Malm J, Tapper U, Tian Y, Penttala V, Karppinen M, Kauppinen E (2009) A novel cement-based hybrid material. New J Physics 11, n° 023013. 10.1088/1367-2630/11/2/023013

NSF (2001) Societal implications of nanoscience and nanotecnology, USA

Pellenq R, Kushima A, Shahsavar R, Vliet K, Buehler M, Yip S, Ulm F (2009) A realistic molecular model of cement hydrates. In: Bazant Z (ed), Northwestern University, PNAS

PICADA (2006) Photocatalytic innovative coverings applications for depollution assessment. innovative facade with de-soiling and de-polluting properties. EC GRD1-2001-00669

Poon C, Cheung E (2007) NO removal efficiency of photocatalytic paving blocks prepared with recycled materials. Constr Build Materials 21:1746–1753. doi:10.1016/j.conbuildmat.2006.05.018

Porro A, Dolado J (2005) Overview of concrete modeling. Proc International conference on applications of nanotechnology in concrete design, pp 35–45

Porro A, Dolado J, Campillo I, Erkizia E, De Miguel Y, De Ybarra Y, Ayuela A (2005) Effects of nanosilica additions on cement pastes. Proc International conference on applications of nanotechnology in concrete design, pp 87–96

Porter D, Vollrath F (2007) Nanoscale toughness of spider silk. Nanotoday 2, 3

Qing Y, Zenan Z, Deyu K, Rongshen C (2008) Influence of nano-SiO_2 addition on properties of hardened cement paste as compared with silica fume. Constr Build Mat 21:539–545. doi:10.1016/j.conbuildmat.2005.09.001

Ramirez A, Demeestere K, De Belie N, Mantyla T, Levanen E (2010) Titanium dioxide coated cementitious materials for air purifying purposes: Preparation, characterization and toluene removal potential. Build Environ 45:832–838. doi:10.1016/j.buildenv.2009.09.003

Reboux G, Bellanger A, Roussel S, Grenouillet F, Millon L (2010) Moulds in dwellings: Health risks and involved species. Rev Mal Respir 27:169–179

Ruot B, Plassais A, Olive F, Guillot L, Bonafous L (2009) TiO_2-containing cement pastes and mortars: Measurements of the photocatalytic efficiency using rhodamine B-based colourimetric test. Sol Energy 83:1794–1801. doi:10.1016/j.solener.2009.05.017

Saepurahman M, Chong F (2010) Preparation and characterization of tungsten-loaded titanium dioxide photocatalyst for enhanced dye degradation. J Hazard Mater 176:451–458. doi:10.1016/j.jhazmat.2009.11.050

Saito T, Iwase J, Horic J, Morioka T (1992) Mode of photocatalytic bactericidal action of powdered semiconductor TiO_2 on mutans streptococci. J Photochem Photobiol B: Biol 14:369–379. doi:10.1016/1011-1344(92)85115-B

Santucci R, Meunier O, Ott M, Herrmann F, Freyd A, De Blay F (2007) Fungic contamination of residence: 10 years assessment of analyses. Rev Franc d'Allergol Immun Clin 47:402–408

Seven O, Dindar B, Aydemir S, Metin D, Ozinel M, Icli S (2004) Solar photocatalytical disinfection of a group of bacteria and fungi aqueous suspensions with TiO_2, ZnO and Sahara desert dust. J Photochem Phototobiol A: Chem 165:103–107. doi:10.1016/j.jphotochem.2004.03.005

Sobolev K, Ferrada-Gutierrez M (2005) How nanotechnology can change the concrete world: part 2. Am Ceram Soc Bull 84: 16–19. www.cognoscibletechnologies.com/.../How-Nanotechnology-Can-Change-the-concrete-world-I.pdf

Sobolev K, Flores I, Hermosillo R, Torres-Martinez L (2008) Nanomaterials and nanotechnology for high-performance cement composites. American Concrete Institute, ACI Special Publication 254: 93–120. https://pantherfile.uwm.edu/sobolev/www/.../7-Sobolev-ACI-F.pdf

Song H, Jiang H, Liu X, Meng G (2006) Efficient degradation of organic pollutant with Wo_x modified nano TiO_2 under visible radiation. J Photochemical Phototobiol A: Chem 181:421–428. doi:10.1016/j.jphotochem.2006.01.001

Strini A, Cassese S, Schiavi L (2005) Measurement of benzene, toluene, ethylbenzene and o-xylene gas phase photodegradation by titanium dioxide dispersed in cementitious materials using a mixed flow reactor. Appl Catal B: Environ 61:90–97. doi:10.1016/j.apcatb.2005.04.009

Taniguchi N (1974) On the basic concept of 'Nano-Technology'. Proc. Intl Conf Prod Eng Tokyo, Part II, Japan Society of Precision Engineering 2:18–23

Vera-Agullo J, Chozas-Ligero V, Portillo-Rico D, Garcia-Casas M, Gutierrez-Martinez A, Mieres-Royo J, Gravalos-Moreno J (2009) Mortar and concrete reinforced with nanomaterials. Nanotechnology in Construction 3, Springer, Berlin Heidelberg

Vohra A, Goswami D, Deshpande D, Block S (2006) Enhanced photocatalytic disinfection of indoor air. Appl Catal B: Environ 65:57–65. doi:10.1016/j.apcatb.2005.10.025

Wang S, Ang H, Tade M (2007) Volatile organic compounds in indoor environment and photocatalytic oxidation: State of the art. Environ Int 42:1843–1850. doi:10.1016/j.envint.2007.02.011

Wiszniewska M, Walusiak-Skorupa J, Gutarowska B, Krakowiak A, Pałczyński C (2009) Is the risk of allergic hypersensitivity to fungi increased by indoor exposure to moulds? Int J Occup Med Environ Health 22:343–354

Yu J (2003) Deactivation and regeneration of environmentally exposed titanium dioxide based products. Testing report, N° E183413, Chinese University of Hong Kong

Zhao J, Yang X (2003) Photocatalytic oxidation for indoor air purification: a literature review. Build Enviro 38:645–654. doi:10.1016/S0360-1323(02)00212-3

Zhu W, Bartos P, Porro A (2004) Application of nanotechnology in construction. Summary of a state-of-the-art report. RILEM TC 197-NCM. Materials Struct 37:649–658. 10.1007/BF02483294

Zyska B (2001) Fungi in indoor air in European Countries. Mikologia Lekarska 8:127–140

Chapter 11
Selection Process

11.1 General

As already stated in Chap. 1, eco-efficient construction and building materials present less environmental impact than common materials. However, it is difficult to say if for instance concrete is more environmentally friendly than steel. Because it is truth that the former is responsible for some CO_2 emissions (0.8 tonnes emitted per tonne produced), on the other hand it uses local raw materials, and may even allow the incorporation of several industrial wastes. The second has the advantage of being recycled indefinitely, but its production involves a higher energy consumption and higher CO_2 emissions (3 tonnes emitted per tonne produced) and is prone to degradation by corrosion. It is then necessary to assess all the environmental impacts of a given material from cradle to grave. This methodology known as "Life Cycle Assesment" (LCA) was used for the first time in the US in 1990. One of the first studies using LCA assess the the resource requirements, emissions and waste caused by different packages of drinks and was conducted by the Midwest Research Institute for the Coca-Cola Company in 1969 (Hunt and Franklin 1996).

11.2 LCA of Construction and Building Materials

The LCA "includes the complete life cycle of the product, process or activity, i.e., the extraction and processing of raw materials, manufacturing, transportation and distribution, use, maintenance, recycling, reuse and final disposal" (Setac 1993). The application of LCA has been regulated internationally since 1996 under ISO14040, ISO14041, ISO14042 and ISO14043. Some of the biggest drawbacks of the LCA, rely on the fact of being very time consuming, implying vast amounts of data on the environmental impacts of materials for all the phases of the life

F. P. Torgal and S. Jalali, *Eco-efficient Construction and Building Materials*,
DOI: 10.1007/978-0-85729-892-8_11,

Table 11.1 Different weightings for categories of environmental impacts (Lippiatt 2002)

Category	Univ. Harvard	EPA
Global warming	6	24
Acidification	22	8
Eutrofication	11	8
Fossil fuel consumption	11	8
Indoor air quality	11	16
Alteration of habitats	6	24
Water consumption	11	4
Air pollutants	228	8

cycle. The categories of environmental impacts commonly used in the LCA, may include the following:

- Consumption of non-renewable resources
- Water consumption
- Global warming potential
- Potential reduction of the ozone layer
- Eutrophication potential
- Acidification potential
- Smog formation potential
- Human toxicity
- Ecological toxicity
- Waste production
- Land use
- Air pollution
- Alteration of habitats.

However, it is understandable that the importance of each category is not the same for each country, being dependent on its environmental specifics. For example a product that consumes a large amount of water, poses a high environmental impact in a very arid country, but that's not the case if the product is produced in a country located in Northern Europe, so it makes perfect sense that the category of environmental impact on drinking water, has a different weight depending on the country where a product or material is produced.

Lippiatt (2002) refers to the case of assigning different categories of environmental impacts by different institutions (Table 11.1).

There are several tools that use LCA and to make an evaluation of the environmental impacts of construction materials such as: BEES (US); BRE. Envest (UK); ATHENA (CANADA); ECOQUANTUM and Simapro (The Netherlands). The software Building for Environmental and Economic Sustainability (BEES), is produced by the US Environmental Protection Agency and is available free of charge to any potential user. BEES has the following impact categories:

- Global warming potential
- Acidification potential

- Eutrophication potential
- Fossil fuel consumption
- Indoor air quality
- Alteration of habitats
- Water consumption
- Air pollutants
- Public health
- Smog formation potential
- Potential reduction of the ozone layer
- Eco-toxicity.

The material performance assessment is made by carbon dioxide units and its contribution for global warming. BEES has a limitation arising from the databases related to US processes, so this tool is recommended only for experimental and educational purposes. The BRE, Envest tool (Anderson and Shiers 2002) uses a notation based on eco-points normalized to the environmental impacts caused by a citizen in the UK during 1 year (100 eco-points). One must bear in mind that the methodologies related to LCA suffer from some uncertainties. In fact it is not possible to tell whether the emission of 1 ton of sulfur dioxide is more polluting than the emission of 3 tons of carbon dioxide or if water pollution is more serious than air pollution, or even if it is possible to know which is the most polluting, the electricity produced by a power plant or by a nuclear power plant. Ekvall et al. (2007) present a more detailed analysis of the LCA limitations. The widespread application of LCA to construction and building materials needs previous surveys on the environmental impacts of these materials throughout their life cycle, something that cannot be extrapolated from studies conducted in other countries due to the different technological and economic contexts.

11.3 Eco-Labels and Environmental Product Declarations

Eco-labels were created to favor the choice of products with enhanced environmental performance and provide a guarantee for a certain environmental performance certified by an independent entity. Since they are quite simple and their meaning is unambiguous these labels have obvious advantages when compared to LCA. Although the advantages of eco-labels are clear, it is important to understand the specifics of the environmental performance in which they are based. Some authors warn that the validity of eco-labels could be in jeopardy if their environmental requirements could be influenced by producer lobbies (West 1995; Ball 2002). On the other hand since the environmental performance of a product or material must include their transportation impacts, there is no way the eco-label can include this impact. So using a particular construction or building material with an eco-label, produced thousands of miles away from the location site, could

Fig. 11.1 Symbol of the German eco-label "Blue Angel"

Fig. 11.2 Symbol of the Canadian "EcoLogo"

be less preferable than the use of local materials, even without that eco-label. Most eco-labels are based on an assessment of the environmental impacts throughout the lifecycle of the product or material in the version "cradle to grave". Germany was the first country to establish in 1978 a labeling system based on environmental criteria with the designation of "Blue Angel" (Fig. 11.1).

Currently, the eco-label "Blue Angel" is applied on 11,500 products covering 90 different categories. This classification means the efficient use of fossil fuels, the reduction of GHG emissions and the reduction of the consumption of non-renewable raw materials, being reviewed every 3 years. The contruction and building materials that already received this label are the following:

- Bituminous coatings
- Bituminous adhesives
- Materials based on glass wastes
- Materials based on paper wastes
- Plywood panels
- External thermal insulation composite systems—ETIC's
- Thermal and acoustic insulation materials
- Wood panels with low VOC emissions
- In 1988 Canada established the label EcoLogoTM (Fig. 11.2), and currently almost 7,000 products are certified by it, including the following construction and building materials:
- Adhesives
- Paints

Fig. 11.3 Symbol of the Nordic eco-label "The Swan"

- Varnishes
- Corrosion inhibitors
- Floor coverings
- Gypsum plaster boards
- Recycled plastic plumbing
- Thermal insulation materials
- Steel for construction.

The use of the EcoLogo implies the respect for a set of environmental procedures dependent on each product. For instance, gypsum boards certified with this label must contain a certain percentage of synthetic gypsum and 100% of recycled paper. In the case of construction steel with the EcoLogo it must contain 50% recycled materials, less than 0.025% of heavy metals and has even to meet a series of environmental requirements during the extraction and production phases. In 1989 the countries of Northern Europe (Finland, Iceland, Norway and Sweden, Denmark only in 1998), created the eco-label "The Swan" (Fig. 11.3). "The Swan" covers 5,000 products of 50 different areas, as below with regard to the construction and building materials area:

- Wood
- Wood panels
- Filling materials
- Materials for floor covering
- Paints and varnishes
- Adhesives
- Windows and doors.

The European "Eco-Label" was created in 1992 (Fig. 11.4), is a system for a voluntary environmental classification for products with low environmental impact throughout its life cycle. The Eco-label applies to a large variety of products with the exception of food, pharmaceutical, medical and hazardous products and like "Blue Angel", involves a periodic review after 3 years. Concerning the construction and building materials, only paints, varnishes and hard floor covering

Fig. 11.4 European Eco-label

materials (tiles, natural stones, concrete, ceramic and clay) are already covered under this label:

- Interior paints and varnishes (2009/544/EC)
- Exterior paints and varnishes (2009/543/EC)
- Hard floor coverings (2002/272/EC; Baldo et al. 2002).

The documents related to the certification of paints and varnishes (Ecobilan 1993) show that its LCA, assessed the following environmental impacts:

- Global warming potential (CO_{eq})
- Potential for atmospheric acidification (increase acidic substances in the lower layers of the atmosphere)
- Eutrophication potential (excess of nutrients from agricultural fertilization)
- Non-renewable resource depletion.

Regarding the hard floor coverings the European Eco-label means that:

- The environmental impacts during the extraction of raw materials were minimized
- During the production phase there is a reduction in overall pollution
- Possible recycled materials were used
- The ceramic tiles are burned with a reduction in the firing temperature.

Eco-labels are advantageous to the final consumer (Kirchoff 2000), however its effectiveness is dependent on the knowledge that consumers may have about their existence and some surveys made in the European Union, indicate that the European eco-label is not well known. In addition to eco-labeling there is another form of environmental certification for construction and building materials known as environmental product declarations (EPDs). They are prepared in accordance with ISO14025 and contain the results of LCA (performed according to ISO14040), of the material or product for the following indicators (Braune et al. 2007):

- Consumption of non-renewable energy
- Consumption of renewable energy
- Global warming potential
- Potential degradation of the ozone layer
- Acidification potential
- Eutrophication potential.

Some authors present information for the development of EPDs for concrete (Askham 2006) and for aluminum (Leroy and Gilmont 2006). An evident disadvantage of EPDs relates to the fact that they do not guarantee a certain level of environmental performance, instead they provide a set of information about it, which only an expert in the field can assess (Manzini et al. 2006; Lim and Park 2009).

11.4 Some Pratical Cases

Several European associations of the concrete industry (BIBM, ERMCO, UEPG, EUROFER, and CEMBUREAU EFCA), in collaboration with the Dutch environmental consultant INTRON BV studied the possibility of minimizing the environmental impacts of concrete elements. One of the objectives of this study, was to develop the tool EcoConcrete, in order to evaluate the environmental impact associated with a particular element of reinforced concrete (Schwartzentruber 2005). Some authors (Gerrilla et al. 2007) compared the performance of houses built with wooden and concrete structures, reporting that the latter had an overall environmental impact only 21% higher than the former. Xing et al. (2008) compared the performance of two office buildings with different structures (reinforced concrete and steel) and found that the steel structure consumes 75% energy compared to the concrete structure and is responsible for half of the emissions GHGs, however, in operational terms the concrete structure exhibits a much lower energy consumption having an overall favorable environmental performance. Marinkovic et al. (2010) studied concretes with and without recycled aggregates and found that their environmental performance is dependent on the transportation distance, regardless of whether they are recycled or not.

Under the project Beddington Zero (Fossil) Energy Development (BedZED), 82 households and 3,000 m^2 of commercial or live/work space with low environmental impact were built in South London. The choice for the construction and building materials in the BEDZED project was made using the BRE. Envest eco-points system (Figs. 11.5, 11.6).

Desarnaulds et al. (2005) also used the BRE. Envest eco-points system to compare different sound insulation materials mentioning that the best environmental performance is associated with recycled paper, followed by rock wool and finally by polystyrene. Nicoletti et al. (2002) showed that ceramic tiles have an environmental impact throughout its life cycle that is over 200% higher than the environmental impact of marble tiles. These results are confirmed by more recent investigations (Traverso et al. 2010). Jonsson (2000) assessed the environmental performance of three floor covering materials using six different approaches:

- An LCA
- An Eco-label (The Swan)
- Two eco-guides (EPM and the Folksam Guide)
- An EPD
- An environmental concept (Natural Step).

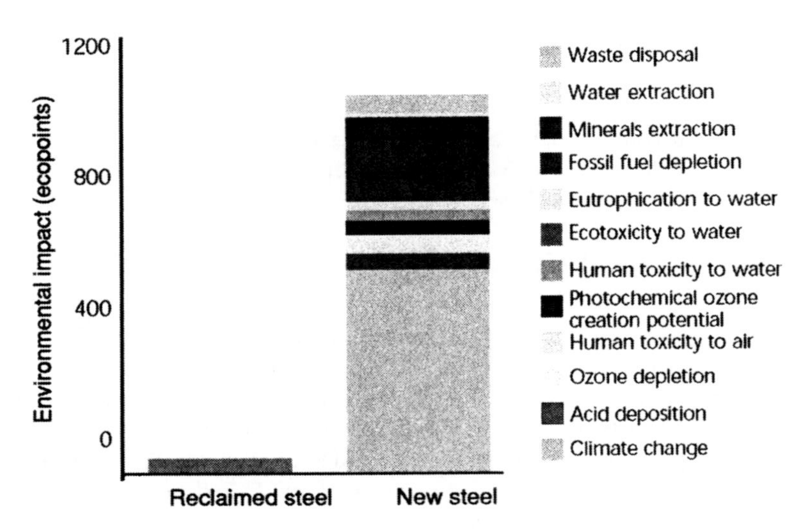

Fig. 11.5 Example of environmental profiling for structural steel (BEDZED2002)

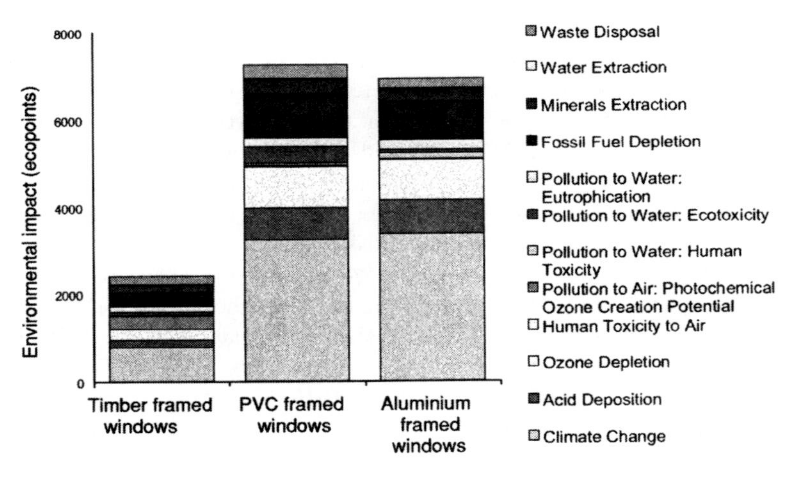

Fig. 11.6 Comparison of the environmental profile of different framed windows (BEDZED2002)

The results showed that while the LCA considers all environmental impacts in a similar way, some forms of sustainability assessment allow prioritizing certain impacts, either during production the phase or during the application of the material in the building. The results also show that only the LCA and the eco-guides allow the development of product rankings. Regarding the aggregation of the results, the eco-label has the best performance and the EPD gets the worst, making it difficult to understand the performance of a particular product.

11.5 Conclusions

Although LCA is the most appropriate way to scientifically evaluate the environmental performance of a given material, it is very time consuming and has some uncertainties. Besides the success of LCA is dependent on the existence (in each country) of lists on the environmental impacts associated with the manufacture of different materials and of the different construction processes. Another drawback of LCA is the fact that it does not take into account possible and future environmental disasters associated with the extraction of raw materials. This means that for instance the LCA of the aluminum produced by the Magyar Aluminum factory, the one responsible in October 2010 for the sludge flood in the town of Kolontar in Hungary, should account for this environmental disaster. Similar considerations can be made about the construction materials that were processed or transported using oil extracted from the Deepwater Horizon well in the Gulf of Mexico. Or even about the materials that were processed using the electricity generated in the Fukushima nuclear power plant. Only then construction and building materials will be associated with their true environmental impact. As for eco-labels they allow a more expedient information for a particular environmental performance, although its value is dependent on the entity and the assumptions that were on the basis of its allocation. Although eco-labels exist for almost 30 years, its use is still neglected by the construction materials market. In fact only a tiny fraction of the current commercial construction materials already have eco-labels. The emphasis in the respect for environmental values will lead to an increase in the number of material producers using eco-labels as a means of differentiation. As regards EPDs they have disadvantages similar to LCA, so it is not expected that in the coming years there may be an accelerated growth of products with EPDs.

References

Anderson J, Shiers D (2002) Green guide to specification. BRE and Environmental profiles, Oxford

Askham N (2006) Excel for calculating EPD data for concrete. In: SETAC Europe 13th LCA case study symposium proceedings with focus on the building and construction sector. Stutgart

Baldo G, Rollino S, Stimmeder G, Fieschi M (2002) The use of LCA to develop eco-label criteria for hard floor coverings on behalf of the European flower. Int J Life Cycle Assess 7:269–275. doi:10.1007/BF02978886

Ball J (2002) Can ISO 14000 and eco-labelling turn the construction industry green. Build Environ 37:421–428. doi:10.1016/S0360-1323(01)00031-2

BEDZED (2002) Bedington zero (fossil) energy development. Construction Materials report. Toolkit for carbon neutral developments-Part 1. BioRegional Development Group http://energy-cities.eu/IMG/pdf/bedzed_construction_materials_report.pdf

Braune A, Kreibig J, Sedlbauer K (2007) The use of EPDs in building assessment—Towards the complete picture. In: Bragança L, Pinheiro M, Jalali S, Mateus R, Amoêda R, Correia Guedes

M (eds) International congress sustainable construction, materials and practices—challenge of the industry for the new millennium, Lisbon.
Desarnaulds V, Costanzo E, Carvalho A, Arlaud B (2005) Sustainability of acoustic materials and acoustic characterization of sustainable materials. Twelth international congress on sound vibration, Lisbon
ECOBILAN (1993) The life cycle, analysis of eleven indoors decorative paints. European Ecolabel, Project for application to paints and varnishes. Ministry of Environment, France
Ekvall T, Assefa G, Bjorklund A, Eriksson O, Finnveden G (2007) What life-cycle assessment does and does not in assessments of waste management. Waste Manag 27:989–996. doi: 10.1016/j.wasman.2007.02.015
Gerrilla G, Teknomo K, Hokao K (2007) An environmental assessment of wood and steel reinforced concrete housing construction. Build Environ 42:2778–2784. doi:10.1016/j.buildenv.2006.07.021
Hunt R, Franklin E (1996) LCA-How it came about. Personal reflections on the origin and the development of LCA in the USA. Int J LCA 1:4–7. doi:10.1007/BF02978624
Jonsson A (2000) Tools and methods for environmental assessment of buildings products—methodological analysis of six selected approaches. Build Environ 35:223–228. doi:10.1016/S0360-1323(99)00016-5
Kirchoff S (2000) Green business and blue angels: a model of voluntary overcompliance with asymmetric information. Environ Resour Economics 15:403–420. doi:10.1016/S0360-1323(99)00016-5
Leroy C, Gilmont B (2006) Developing and EPD tool for aluminium building products: the experience of the European aluminium industry. In: SETAC Europe 13th LCA case study symposium proceedings with focus on the building and construction sector. Stuttgart
Lim S, Park J (2009) Environmental indicators for communication of life cycle impact assessment results and their applications. J Environ Manag 90:3305–3312. doi:10.1016/j.jenvman.2009.05.003
Lippiatt B (2002) BEES®3.0 Building for environmental and economic sustainability technical manual and user guide. National Institute of Standards and Technology
Manzini R, Noci G, Ostinelli M, Pizzurno E (2006) Assessing environmental product declaration opportunities: a reference framework. Bus Strategy Environ 15:118–134. doi:10.1002/bse.453
Marinkovic S, Radonjanin V, Malesev M, Ignjatovic I (2010) Comparative environmental assessment of natural and recycled aggregate concrete. Waste Manag 30:2255–2264. doi: 10.1016/j.wasman.2010.04.012
Nicoletti G, Notarnicola B, Tassielli G (2002) Comparative life cycle assessment of flooring materials: ceramic versus marble tiles. J Clean Prod 10:283–296. doi:10.1016/S0959-6526(01)00028-2
Schwartzentruber A (2005) EcoConcrete: A tool to promote life cycle thinking for concrete applications. Orgagec symposium 2–10
SETAC (1993) Society of environmental toxicology and chemistry–guidelines for life-cycle assessment: a code of practice. Elsevier, Brussels
Traverso M, Rizzo G, Finkbeiner M (2010) Environmental performance of building materials: life cycle assessment of a typical Sicilian marble. Int J Life Cycle Assess. 15:104–114. doi: 10.1007/s11367-009-0135-z
West K (1995) Eco-labels: the industrialization of environmental standards. Ecologist 25:31–47
Xing S, Xu Z, Jun G (2008) Inventory analysis of LCA on steel- and concrete-construction office buildings. Energy Build 40:1188–1193. doi:10.1016/j.enbuild.2007.10.016

Subject Index

F. P. Torgal and S. Jalali, *Eco-efficient Construction and Building Materials*,
DOI: 10.1007/978-0-85729-892-8, © Springer-Verlag London Limited

X

Lightning Source UK Ltd.
Milton Keynes UK
UKOW021358040212

186675UK00001B/5/P

Ollscoil na hÉireann, Gaillimh
3 1111 40257 7496